D1135291

Men, Women and War

T.K. DANIEL 1927–1990

Men, Women and War

Edited by T.G. Fraser and Keith Jeffery

HISTORICAL STUDIES XVIII

Papers read before the xxth Irish Conference of Historians,
held at Magee College, University of Ulster, 6–8 June 1991

Thomas Bartlett
Ida Blom
Irene Collins
T.G. Fraser
John Horne
Keith Jeffery
Peter Lowe
Brian Manning
Eunan O'Halpin
Ruth Roach Pierson
J.R.S. Phillips
René Pillorget
Stephen G. Rabe
Jean Stengers

THE LILLIPUT PRESS

First published in 1993 by
THE LILLIPUT PRESS LTD
4 Rosemount Terrace, Arbour Hill,
Dublin 7, Ireland.

A CIP record for this
title is available from
The British Library.

ISBN 0 946640 96 3

Jacket design by Jarlath Hayes
Set in 10 on 12.5 Centaur by
mermaid turbulence
Printed in Dublin by ßetaprint

CONTENTS

ILLUSTRATIONS

between pp. 132–3

Cover of booklet produced in 1915 for primary schoolchildren (Bibliothek für Zeitgeschichte, Stuttgart)

Lucien Jonas, 'Le salut au blessé', *L'Illustration*, 20 February 1915

between pp. 154–5

William Orpen, 'To the Unknown British Soldier in France' (original version) (Imperial War Museum, London)

William Orpen, 'To the Unknown British Soldier in France' (final version) (Imperial War Museum, London)

William Conor, from a fund-raising booklet in aid of the Ulster Volunteer Force Hospital (by permission of the Linen Hall Library, Belfast)

p. 164

Figure: Military Intelligence, Second Bureau, 1925

p. 176

Map: the situation on 24 May 1940

between pp. 208–9

Canadian Women's Army Corps recruiting stand (Manitoba Archives, Canadian Army Photograph collection, no. 162)

Female armaments worker (City of Toronto Archives, G&M 91963)

Army driver (Manitoba Archives, Canadian Army Photograph collection, no. 157)

Women workers' Beauty Contest (City of Toronto Archives, G&M 80371)

FOREWORD

With the exception of the introductory chapter, the papers in this volume were presented at the xxth Irish Conference of Historians, held at the University of Ulster Magee College on 6–8 June 1991. This biennial conference, run under the aegis of the Irish Committee of Historical Sciences (the representative body for professional historians throughout the island of Ireland), is the chief general meeting of academic historians in Ireland, and the aim of the occasion has always been to combine scholarly perspectives from both Ireland and elsewhere, and also to cover as wide a chronological range as possible.

The conference was organised by the University of Ulster History Department and we are very grateful to the following for financial and other assistance in mounting the event: the Northern Ireland Department of Education; the University of Ulster Faculty of Humanities; the Provost of Magee College; the Garfield Weston Trust; the British Academy; the Honourable the Irish Society; the Royal Belgian Benevolent Society; the French Embassy; the Royal Norwegian Embassy; and Derry City Council.

In the preparation of the papers for publication we owe a particular debt of gratitude to Joanne Taggart, who has meticulously typed up the text; Tony Feenan, who helped prepare the illustrations; and Gillian Coward, who drew the map. Antony Farrell and Mari-aymone Djeribi of The Lilliput Press have greatly eased the burdens of the editors with their expert assistance and encouragement. We would also like to acknowledge permission to reproduce illustrations from the following institutions: the Bibliothek für Zeitgeschichte, Stuttgart; the Linen Hall Library, Belfast; the Imperial War Museum, London; the Manitoba Archives Canadian Army Photograph Collection; and the City of Toronto Archives.

The publication of these proceedings has been made possible through generous grants from the University of Ulster Academic Publications Committee and from the estate of the late Mr T.K. Daniel. Mr Daniel, a mature student, was among the very first cohort of history students to pass through the New University of Ulster after its foundation in 1968. Subsequently he became a lecturer in history at the Ulster Polytechnic, which merged with the New University in 1984 to form the present University of Ulster. He was an inveterate and enthusiastic conference-goer, and, had he not died in December 1990, would undoubtedly have made a significant contribution to the proceedings of the conference at Magee College. This book is dedicated to his memory.

T.G. Fraser and Keith Jeffery
University of Ulster, May 1993

T.D. Williams (ed.), *Historical Studies I* (London: Bowes and Bowes 1958)

M. Roberts (ed.), *Historical Studies II* (London: Bowes and Bowes 1959)

J. Hogan (ed.), *Historical Studies III* (London: Bowes and Bowes 1961)

G.A. Hayes-McCoy (ed.), *Historical Studies IV* (London: Bowes and Bowes 1963)

J.L. McCracken (ed.), *Historical Studies V* (London: Bowes and Bowes 1965)

T.W. Moody (ed.), *Historical Studies VI* (London: Routledge & Kegan Paul 1968)

J.C. Beckett (ed.), *Historical Studies VII* (London: Routledge & Kegan Paul 1969)

T.D. Williams (ed.), *Historical Studies VIII* (Dublin: Gill & Macmillan 1971)

J.G. Barry (ed.), *Historical Studies IX* (Belfast: Blackstaff Press 1974)

G.A. Hayes-McCoy (ed.), *Historical Studies X* (Dublin: ICHS 1976)

T.W. Moody (ed.), *Nationality and the Pursuit of National Independence: Historical Studies XI* (Belfast: Appletree Press 1978)

A.C. Hepburn, (ed.), *Minorities in History: Historical Studies XII* (London: Edward Arnold 1978)

D.W. Harkness & M. O'Dowd (eds), *The Town in Ireland: Historical Studies XIII* (Belfast: Appletree Press 1981)

J.I. McGuire & A. Cosgrove (eds), *Parliament and Community: Historical Studies XIV* (Belfast: Appletree Press 1985)

P.J. Corish (ed.), *Radicals, Rebels and Establishments: Historical Studies XV* (Belfast: Appletree Press 1985)

Tom Dunne (ed.), *The Writer as Witness: Literature as Historical Evidence: Historical Studies XVI* (Cork: Cork University Press 1987)

Ciaran Brady (ed.), *Ideology and the Historians: Historical Studies XVII* (Dublin: The Lilliput Press 1991)

CONTRIBUTORS

Thomas Bartlett
Lecturer in History, University College Galway
Ida Blom
*Professor of History, University of Bergen, and President of the International
Federation for Research in Women's History*
Irene Collins
Emeritus Reader, University of Liverpool
T.G. Fraser
Professor of History, University of Ulster
John Horne
Lecturer in Modern History and Director of European Studies, Trinity College Dublin
Keith Jeffery
Senior Lecturer in History, University of Ulster
Peter Lowe
Reader in History, University of Manchester
Brian Manning
Emeritus Professor of History, University of Ulster
Eunan O'Halpin
Senior Lecturer in Government, Dublin City University
Ruth Roach Pierson
*Professor of Women's History and Feminist Studies, Ontario
Institute for Studies in Education, Toronto*
J.R.S. Phillips
Professor of Medieval History, University College Dublin
René Pillorget
Emeritus Professor, University of Lille
Stephen G. Rabe
Professor of History, University of Texas at Dallas
Jean Stengers
*Professor of Contemporary History, University of Brussels,
and Member of the Belgian Royal Academy*

MEN, WOMEN AND WAR

T.G. Fraser

War, William Tecumseh Sherman famously observed, is Hell, but it remains a subject of abiding fascination to the public and the professional historian alike. In 1933 David Lloyd George complained that his 'library shelves groaned under the burden of war autobiographies'.¹ What might he say now, almost sixty years later, of the libraries of books about the war he helped wage or about the even greater world conflict that followed? There is no lack of issues to fire the controversies and debates which carry forward the study of history. Old questions remain. Why did Lee reject the advice of his subordinates and launch the disastrous charge on the Union lines at Gettysburg? Why did Bluecher's Prussians and not Marshal Grouchy's two corps arrive at the critical moment on the field of Waterloo? Such issues, which have long vexed historians, are still relevant, for a successful invasion of Pennsylvania could perhaps have confirmed the South's bid for independence, with incalculable consequences for America's rise as a world power; the defeat of Wellington's army could either have forced monarchical Europe to accept the restoration of Napoleon's empire or have ushered in yet another period of continental war. But new questions have arisen. Has war, for example, with its preoccupation with the male warrior figure, been instrumental in preserving masculine hegemony in society? Have the two great wars of this century really done anything to help reverse, or even mitigate, that imbalance in Western society? This volume, ranging as it does from the Middle Ages in Ireland to America's débâcle in Vietnam, cannot pretend to examine, still less reassess, all the complex issues raised by war, but it is an attempt by an international group of specialists to offer fresh insights into an area of abiding interest and concern.

The aspect which most attracted earlier generations of historians was the conduct of war. Although this has ceased exclusively to be at the cutting edge of the discipline, and no one advocates returning to the 'drum and trumpet' manner of historical writing, historians realise that to write about war with the fighting taken out is a distortion of the past. The innate conservatism in

British military circles, itself a reflection of society, which balked at the idea of the tank replacing the well-bred horse and the mechanic replacing the cavalryman, explains the disasters which befell the Eighth Army in June 1942, but what of the lonely decisions taken by General Auchinleck and his staff as they contemplated the wreckage of their army along the Alamein positions? Theirs was not the leisure to contemplate the shortcomings of British society. They could not stop to analyse the ways in which German generals had rewritten the rules of armoured warfare in the 1930s. Their task was to make hasty plans with what human and material resources remained to them. In so doing, they halted the advance of Rommel's army and permanently changed the course of the war in North Africa. Two years before, Rommel and his fellow panzer generals had destroyed French and British power in Europe. What sustained Britain through this hour of humiliation was the safe return of the bulk of the British Expeditionary Force. Although, as Churchill reminded his countrymen, wars are not won by evacuations, it is difficult to see how resistance to Hitler's ambitions could have continued but for the 'miracle of Dunkirk' with its legends of the little ships snatching the troops from the victorious enemy. Without attempting to belittle that achievement, the truth is more complex, for it was a miracle that should never have been permitted. The facts seem simple enough. On 24 May, after one of the most dramatic campaigns in history, the German panzer divisions which were ready to destroy the retreating British army around Dunkirk were halted on the Führer's express order. By the time that order was rescinded two days later, the British had consolidated their positions in such a manner as to permit 'The Deliverance of Dunkirk'.[2] When the evacuation ended on 4 June, 338,226 British and French soldiers had been saved. The *Haltbefehl* was one of the most fateful decisions of the Second World War, and one of the most controversial. When one dismisses the fanciful idea that the order was a political gesture by Hitler to encourage the British to make peace, a variety of military possibilities remains. These are examined in detail by Jean Stengers who agrees with those analysts who point to Hitler's desire to conserve his armoured forces for the final battle against France, and to allow Goering's Luftwaffe the final credit for Britain's defeat.[3] On 27 May, with the ill-informed exuberance that was to cost Germany so dear in the course of the war, Goering informed his master that 'Nur Fischkutter Kommen herüber; hoffentlich können die Tommys gut schwimmen'. To these reasons, Stengers suggests another; namely, that for political reasons Hitler wished to spare the Flemish areas of Belgium. Throughout his career Hitler had an agenda of racial conquest independent of, and generally concealed from, the profession-

als of the Foreign Ministry and the General Staff. As part of that, he envisaged the 'Germanic' elements in the European population welded to the core
of the Third Reich: the Flemish fitted this stereotype. Racial blindness and
prejudice led him to some of the greatest crimes in history. Stengers argues it
also contributed to his ultimate defeat, for in trying to spare Flanders the full
fury of his blitzkrieg he enabled Britain to survive.

The Second World War failed to conform to Hitler's preconceptions.
Wars seldom fulfil their authors' expectations. Britain did not, after all,
respond to his peace overtures, nor did Soviet Union collapse in six weeks,
and by 1942 Germany was facing an acute problem of recruitment. Turning at
first to the *Volksdeutsch* of eastern Europe, the German leadership then had to
enlist such volunteers as they could raise in 'Germanic' Europe, Scandinavia
and the Low Countries, and, finally, in defiance of their own racialist notions,
to the Ukraine, the Baltic lands, the Balkans, and to anyone attracted by Nazi
ideology. It is one of history's finer ironies that prominent amongst the final
defenders of Hitler's Chancellery were French SS troopers. Such a bizarre
twist of fate simply points to the age-old problem of recruitment. It certainly
faced Napoleon, and for similar reasons. Just as Hitler over-extended the
Third Reich's capacity by engaging the Soviet Union, the United States and
the British Empire, the French Emperor found himself trying to sustain a war
effort which by 1812 stretched from the Iberian Peninsula to Moscow. It
could not be done. 'In 1813', as Irene Collins points out, 'rules would go by the
board – previous classes would be combed out, future classes anticipated, and
men of less than regulation height sent to the front.' Not only did this signal
the crisis of the Empire, it turned popular opinion against it. Unavailable to
Napoleon, of course, was the mobilisation of women, so striking a feature of
the twentieth century. It was less true of Nazi Germany, where Hitler preferred as far as possible to relegate women to the *Kinder, Küche, Kirche* role,
using instead foreign slave labour, but in the Western democracies, and in the
USSR, female recruitment into the armed forces and the war economies made
an inestimable contribution to victory. 'Rosie the Rivetter' became a folk
heroine in the United States, while in Canada 50,000 women served in the
armed forces and one million in war production.[4] 'Womanpower', observes
Ruth Roach Pierson, was used to relieve Canada's 'manpower' problems.

Such shortages had earlier curbed Britain's efforts during her long series of
wars with France. Just as twentieth-century governments were forced to turn
to women, so British governments had to overcome their prejudices against
recruiting Irish Catholics, and with much the same kind of ambivalence. But,
as Thomas Bartlett explains, a war machine which expanded from 30,000 in

the early eighteenth century to nearly 900,000 in 1809 had no choice. By 1830 some 40 per cent of the British Army was Irish. Such devices may be needed to win wars but raise awkward questions for those who resort to them. Would women used to independence and authority in war conform to accepted gender stereotypes once it had ended? Troops from Catholic Ireland might, as Lord Sidmouth conceded, have 'turn'd the scale on the 18th June at Waterloo', but what would they expect in return – the Ireland of 'Papists lie down'?

With the French Revolution, wars ceased to be the preserve of the well-drilled professional, becoming instead the concern of mass societies. The revolutionaries understood this with their proclamation of the *levée en masse*. Slowly, and only after many painful defeats at the hands of the new French armies, did the rest of Europe grasp what was needed to challenge Napoleon's hegemony. 'Nun Volk steh' auf und Sturm brich los!', the battle-cry of 1813, would have been incomprehensible to the Prussians of Jena in 1806.[5] Napoleon, of course, well understood the need to associate the nation with his military campaigns. Hence, Irene Collins explains, the dressing of his legislators in quasi-military uniforms, the ceremonial presentation of captured flags to parliament and his attempt to convert the Madeleine into a Temple of Glory. A century later, with the creation of a mass society in the modern sense, questions of public opinion and morale were to preoccupy all governments in the First World War, not least the French. Lingering sentiments of Napoleonic Glory were hard to reconcile with the miseries of the Chemin des Dames, forcing the government to conjure up fresh images to sustain the unity of soldier and civilian.[6] John Horne's chapter examines the nature of French wartime propaganda which resulted. 'The state's preoccupation', he notes, 'with public opinion and morale is the surest indication that the more or less willing commitment of the action and energy of the belligerent populations was vital for prosecuting the wars.' Such propaganda must idealise one's own side, 'the hirsute soldier hero of the Marne', and demonise the enemy. For some, the legacy was not as straightforward. The war memorials of Alsace are a study in ambiguity, dedicated 'A nos victimes de guerre' rather than 'Mort pour la patrie'. For a generation Hungarians bridled at the number of Soviet war memorials in their country, while nothing commemorated their own dead of World War Two. The Soviet soldier at the base of the Liberation Monument on Budapest's Gellert Hill has now been removed. Austrians, too, faced an uncomfortable legacy after 1945. Hence the memorial to the 9th Panzer Division in Vienna's garrison church refers to its earlier existence before the *Anschluss* and has ribbons in both the Austrian and

German national colours. The memorial to Colonel-General Alexander Löhr, Austrian Commander of Army Group E in the Balkans, almost caused a breach between the Austrians and Yugoslavs, who had executed him for war crimes.[7]

And what of the Irish, of whom 35,000 died in the Great War? Keith Jeffery's paper exposes the war's tortured legacy for both parts of Ireland. War memorials erected in the Irish Free State had to be as delicately inscribed as those of Alsace, albeit for rather different reasons. Politicians and veterans in newly independent Ireland wrestled for nearly two decades with the problem of a national memorial to their fellow-countrymen, only the deteriorating situation in 1939 enabling them to avoid a formal opening of the memorial park at the 'distant backwater' of Islandbridge. Loyalist Ulster saw things differently. 'The war, and the battle of the Somme especially', argues Jeffery, 'were presented as a Unionist blood sacrifice: the Union permanently sealed with blood.' An understandable position, it did little for the memory of West Belfast Catholics who had died with the 16th Irish Division. The way the war was remembered not only underlined the growing separateness of North and South; it also emphasised the gulf inside Northern Ireland between those who claimed for themselves the legacy of the war and their political opponents.

By 1939 propaganda had assumed an even more central position in war. The development of the radio and of cinema newsreels had seen to that. In 1933, Josef Goebbels had become Minister of Propaganda and Public Enlightenment in the new National Socialist State. It was a measure of his importance that as the war developed so did his role. By 1943 he was Plenipotentiary for Total War and as Hitler retreated from the realities of the struggle he had unleashed so Goebbels came to represent to the German public the spirit of resistance. But the allies, too, were fully aware of the need for war propaganda. Ruth Roach Pierson's chapter illustrates how the sexuality of Canadian women was exploited for war purposes. 'Munitions, armament and aircraft plants staged annual "beauty contests" to raise the morale of their women workers and to parade the company's patriotism before the public', she relates. Such things are understandable when there is a broad consensus that war is justified but where that is not the case then the relationship between war and society becomes problematic at best. Nowhere has this been better illustrated than over the Vietnam War. Defeat in south-east Asia has cut a deep scar in American society, most obviously in the so-called 'Vietnam generation'. A member of that generation, Stephen Rabe has contributed a courageous and perceptive example of contemporary history. Like many of his peers, Rabe declares, 'I thought then and I believe now that intervention in

Vietnam was not in the best interests of the United States'. He has, however, been forced to confront an increasingly influential school of revisionism which has argued that failure was the result of domestic criticisms so violent that the hands of the executive, Congress and the military command were fatally tied.[8] This charge, Rabe believes, cannot be sustained by the statistics of the war. With 50,000 American and two million Vietnamese dead, 'how much more blood and treasure', he asks 'should have been expended?' What concerns him is that revisionists have been asking the wrong question. The key to Vietnam is to understand 'why the Vietnamese communists withstood US onslaught'. The United States was too powerful ever to be defeated, but the *Viet Cong* and the North Vietnamese were still able to deny her victory. It is the complex legacy of that reality which unites the 50,000 names on the long wall in Washington with those of the four students on the poignant memorial in the grounds of Kent State University who were shot by the National Guard whilst protesting against America's role in the war.

Rabe's paper serves to remind us of the truism that wars generally result from miscalculation and misperception, Bismarck's claims about the wars of German unification notwithstanding.[9] In 1939 Hitler failed to perceive how the British and French would regard his planned moves against Poland and hence miscalculated their reaction. In that he was no wiser than his imperial predecessors in Berlin and Vienna in 1914. War brutally exposes such miscalculations. Peter Lowe's paper explains how Britain tried to foster a positive relationship with the new communist regime in China which came to power in 1949 after defeating the *Kuomintang* in the civil war. There were good reasons for doing so. The *Kuomintang* had been venal and incompetent, while the communists were clearly in charge of the country. British recognition of the new government would, it was hoped, preserve Britain's commercial position while signalling good intentions towards the new forces at work in post-war Asia. Understandable though this position was, Lowe demonstrates how it was confounded by the complexities of the situation in East Asia and in American politics. These came together in the Korean War which found British troops fighting alongside the Americans against the Chinese who had come to the aid of the North Koreans. Just as the Vietnam war destroyed America's pretensions to globalism, Korea confounded British hopes of turning the revolution in China to her advantage. Given the weakness of her position and the strongly anti-imperialist nature of the Chinese regime, it had never been a realistic perception, as the war cruelly showed.

Precisely because it strips away illusions, war accelerates the pace of political change. Men and women who have undergone the experience of war are rarely content with the *status quo ante*, as Churchill discovered to his sorrow in 1945. Discontented soldiers dismissed him as effectively, if less bloodily, as their Russian counterparts had done the Romanovs in 1917. Since 1914 war and revolution have been closely linked but this is by no means unique to this century. Brian Manning's chapter addresses the question of whether events in the New Model Army in 1648–9 were a military *coup d'état* or whether the soldiers had become a revolutionary force. It is an important question for our understanding of the English Revolution. The officers of the New Model Army had enlisted to fight for principles and in the main they did not come from the established élite. The officer who arrested King Charles in 1647 had been a tailor. Hence, their activities carried with them the hope of social as well as of political revolution. Manning's careful analysis of the army mutinies of 1649, examining the extent of radical influence and of grievances over pay or service in Ireland, points to a critical juncture in the English Revolution. He concludes that the loyalty of the soldiers was shaken but not shattered: 'The mutinies of 1649 were a watershed. Their suppression was a victory for military force and a defeat for the more revolutionary and populist section of the "middle sort of people". In the late 1640s the army was a revolutionary force: in the 1650s it became a force for order and stability, inhibiting or repressing radical dissent and suppressing popular resistance.' In short, the army had decided the nature and extent of the Revolution. Just as the English Revolution became transformed into the rule of Oliver Cromwell, so the French Revolution was taken over by its ablest soldier. Many, most obviously Edmund Burke, had from the start predicted just such an outcome and as the new Republic battled against the assembled forces of monarchical Europe it was inevitable that its leaders would turn to a military man to ensure its survival. It was by no means inevitable, however, that this new Caesar would be Napoleon Bonaparte. While some established commanders like Massena and Jourdan were content to remain soldiers, others, Moreau, Bernadotte and Hoche, combined substantial military reputations with some political ambition. In the end, of course, it was Napoleon who possessed the necessary combination of military renown, capacity for intrigue and the right political connexions which enabled him to grasp power. But as he progressed from Consul to Emperor and then to master of much of Europe, Napoleon was concerned to be seen to act within a constitutional framework. It was only in the campaign to defend France in 1814 that he brazenly defied the constitution. This was a sign that the Empire had run its course, for in its heyday, as Irene Collins shows, Napoleon had consciously tried to associate his parliaments with the benefits and glories of his rule. After the stimulation of revolution and military rule came reaction. Just as Cromwell's death saw the restoration of the Stuarts, Napoleon's defeat was followed by the return of the Bourbons. After a brief flirtation with the politics of accommodation, each dynasty reverted to type and was soon dismissed, this time for good.

In the old whig view, the removal of James II secured English liberty, or at least a version of it. But his defeat at the Boyne confirmed the political eclipse of Irish Catholicism.[10] In the early eighteenth century, as Bartlett acknowledges, to be Catholic was to be identified as Jacobite and hostile to the Williamite and Hanoverian settlements. After 1745 Catholics were able to demonstrate their loyalty through military service, thus posing an inescapable constitutional question against their continuing disabilities. War also enabled women to challenge their exclusion from political power. The significance of the First World War for female franchise in Britain is well known and in Scandinavia, as Ida Blom shows: 'female responses to defence policy and activity during the final national crisis in 1905, earned women a sympathetic reaction when the question of the vote was discussed in Parliament and contributed to victory in 1907.' Such interactions are not unique to the twentieth century or the modern period. In the Middle Ages, war could result not just in a change of ruler but in the transformation of an entire governing élite. By the time Domesday Book was compiled not only had the great Anglo-Norse earldoms disappeared but there were only two non-Norman landholders left in England, such was the impact of the Norman Conquest. 'By about 1100 no English secular or ecclesiastical lord of any importance still held power, while the work of the government was conducted exclusively in either Latin or French', Seymour Phillips observes. Phillips is concerned to investigate the continuing implications of these revolutionary events in the early fourteenth century when Ireland, itself colonised by the Normans, was invaded by the scion of another great Norman family, Edward Bruce. Had Bruce's ambitions not been thwarted, there could have been the most profound restructuring of relationships, for his brother was well on the way to unimpeded authority in Scotland as the result of his decisive victory at Bannockburn. Even so, the period of his invasion of Ireland is one which is critical for our understanding of Norman and Gaelic Ireland for the failure of Bruce and his Irish allies ensured that the 'English-ruled lordship of Ireland ... staggered on until 1541 when it was finally extinguished at the hands of the King of England himself.'

There remains the intriguing question of the relationship between war and society at large. When the leaders of the southern states declared for secession from the United States in 1861, it was partly in the belief that their inferiority in numbers and resources would be offset by the reluctance of the northern population to sustain a prolonged struggle. Despite the débâcle of First Bull Run, McClellan's humiliation on the Peninsula, Pope's fatuous performance at Second Bull Run, the slaughter of Antietam, the awful carnage at Fredericksburg, and Lee's brilliant victory at Chancellorsville, they were wrong. When General Meade's line held at Gettysburg, the Union was saved. Eighty years later, the rulers of Imperial Japan made a similar fateful calculation about American resolve and were equally confounded. If a defensive perimeter were created, they believed, American public opinion would never sustain the casualties needed to penetrate it. Despite the ferocious heroism of Japanese forces, Guadalcanal, Rabaul, the Marianas, the Carolines, Leyte Gulf, Iwo Jima and Okinawa proved that American society was equal to the challenge they had thrown down. 'Remember Pearl Harbor' proved to be no empty boast. The Second World War saw the identification of modern mass society with the war effort and it was a competition in which democracy in alliance with Stalinism won triumphantly over dictatorship. Canada's women rallied to their country's war effort in a way which Germany's leaders never dared to encourage, at least until it was too late. Churchill's appeal to 'blood, sweat, toil and tears' after Dunkirk was only echoed in Germany after the disaster at Stalingrad in 1943. Even then, Hitler dared not match Churchill. The carefully nurtured 'Hitler Myth' could not survive defeat. It fell to Goebbels to ask the party faithful at the Berlin Sportpalast if they wanted total war, too late to be any use." As Rabe's paper confirms, society's support is vital if a war is to succeed. Ordinary men and women must give their consent and then be willing to make the sacrifices. Napoleon had forfeited that consent by 1813. So had Lyndon Johnson and Richard Nixon over Vietnam, at least for a significant number of people. The *Viet Cong* had not and won.

Finally, war brings into sharp focus the role of the armed forces in society. Armaments are not cheap. By 1813, French parliamentarians balked at paying the human and financial cost of Napoleon's wars. The United States poured a fortune into the Vietnam War. When an army is clearly defending national territory against an aggressor, then its demands on resources are rarely an issue. The French army of 1914 was seen in just such a way, though perceptions changed as the war dragged on to no obvious conclusion. But armed forces can represent repression as easily as defence. René Pillorget demonstrates how royal troops were used to suppress dissent in the Midi in the early seventeenth century. The soldiers of Louis XIII and Richelieu regarded the Protestant areas of Languedoc as enemy territory. 'When the pious King Louis XIII took Privas, he had a hundred of the principal prisoners hanged without a trial and sent a hundred more to the galleys'; their purpose Pillorget concludes, being 'to terrorise the Protestant population by an act of exceptional severity'.

The purpose of a military establishment is the theme explored by Eunan O'Halpin. Like the other small states of inter-war Europe, Ireland could not hope to deploy a military force capable of defending itself against a powerful enemy, but like most of them it was vulnerable to internal threat. The Austrian army could not have stopped Hitler, even if it had wanted to, but it was remarkably effective against the Viennese workers in 1934. O'Halpin's analysis confirms this for Ireland. Concluding that no Irish government 'has been willing to endorse a remotely credible policy of external defence', he argues that the army 'has been highly effective in defending the state from within.' His sensitive analysis of the ways in which the Irish army steered a course through the violent early years of the state and the dangers of the Second World War confirms the pivotal role which military intelligence often plays in shaping policy. The authorities in Dublin were well served.

While military force is able to preserve state hegemony, it can also be argued, that war has confirmed male domination of society. Armies are aggressively male organisations in which women have traditionally played a peripheral part. The Great Commanders from Alexander of Macedon to the present have been men operating in an almost exclusively masculine world, some of them, like Prince Eugene or Frederick the Great, openly misogynist. Women might attract romantic loyalty: the Hungarian nobles drawing their swords in defence of Maria Theresa in 1741 is a notable example, even though they only did so after Hungary's rights had been confirmed.[12] At worst, women might be no more than 'camp followers': what more pejorative term could there be than *Feldhure*? At best, they might fulfil women's traditional stereotype as the nurturer of life: witness the largely unsung heroism of the Sanitary Commission in the American Civil War. Union General Francis Barlow survived appalling wounds at Antietam and Gettysburg; his wife Arabella died of typhus tending the wounded at the siege of Petersburg.[13] Women could symbolise what men were protecting. The imagery of the First World War has the chaste figures of 'Germania', 'Austria', 'Britannia', 'Hibernia' and 'Columbia' representing their countries. But by then women were starting their challenge. As Ida Blom shows, this had been stirring even before 1914. Scandinavian women took up the prospect of war to assert a new role. It was no accident that 'their' warship was named *Valkyrien*, for these were the warrior-maidens of Norse antiquity. By the Second World War women were taking their full part, or almost so, for, as Ruth Roach Pierson argues, Canadian women were confined to supporting the fighting front: 'The Girls behind the Boys behind the Guns'. The convention that only men are suited to the dangers of combat could not survive the merest acquaintance with the careers and fates of women like Odette Churchill or Violette Szabo, but it is a stubborn one. The role of women in the fighting front may yet prove to be the acid test of equality.

The following chapters explore these themes in detail, analysing issues which, though often uncomfortable, are inescapable for any student of history. Western Europe's freedom from war since 1945 gives no cause for complacency, for Europe also knew a long period of peace after 1815. The conference which discussed these issues was held in an historic European city where the siege of 1688–9 still finds an echo. It took place only weeks after war in the Gulf and as the disintegration of Yugoslavia ushered in new levels of ethnic violence. This book offers neither solutions nor reassurance, only the attempt of a group of historians to find fresh insights into the nature of war and the place of men and women in it. The men and women – and children – of Vukovar and Sarajevo in 1993 would have known what General Sherman meant.

NOTES

1   David Lloyd George, *War Memoirs*, new edition, vol. i (London 1938), p. vii.

2   See Winston C. Churchill, *The Second World War*, vol. ii, *Their Finest Hour* (London 1949), chapter v, 'The Deliverance of Dunkirk'. His 1940 speeches still reward analysis.

3   For a recent appraisal of Hitler's motives, see Alan Bullock, *Hitler and Stalin Parallel Lives* (London 1991), p. 740.

4   Rosie's slogan 'We can do it' became celebrated in film and on poster and has, for obvious reasons, enjoyed something of a revival.

5   In 1943 Goebbels revived the old slogan of 1813 to animate Germany's resistance, without notable success. In 1945, the aged and poorly armed *Volksturm* generally had more sense than to take on the Allied armies.

6   The traditions of Napoleonic glory were embodied in 'Plan 17'. Completed in 1913, its purpose was to sweep the French armies across the Rhine with irresistible dash. The result should have been predictable. For an analysis, see Barbara Tuchman, *August 1914* (London 1962).

7   Löhr insisted on accompanying his men into Yugoslav captivity. Unable, it would seem, to find him guilty of specific war crimes, the Yugoslavs resorted to condemning him for having been responsible for planning the Luftwaffe's attack on Belgrade in 1941 in which some 15,000 died. Many of those who served in Yugoslavia were, like Löhr, Austrians whose feelings towards the Serbs could be assumed. Kurt Waldheim's election as President of Austria in 1986 revived these issues for a time. See R.E. Hertzstein, *Waldheim, the Missing Years* (London 1988).

8   A powerful statement of this thesis is, predictably, Richard Nixon, *No More Vietnams* (London 1985). Operation 'Desert Storm' against Iraq in 1991 was held by many to have exorcised this ghost, even though Saddam Hussein's regime survived.

9   Bismarck's *Recollections and Reminiscences* (2 vols, London 1898), is perhaps the classic state-
     ment of how events unfolded according to the author's plans, or could be made to seem
     so.

10   See S.J. Connolly, *Religion, Law and Power: the Making of Protestant Ireland 1660-1760* (Oxford
     1992).

11   See Ian Kershaw, *The Hitler Myth* (London 1987). Even so, the 'Myth' seems to have
     remained remarkably potent, people preferring to blame the incompetence of the Party,
     or the Luftwaffe, for the country's disastrous position.

12   Johann Strauss's opera *Der Zigeunerbaron* charmingly recreates what was, in fact, a tempes-
     tuous relationship between the Hungarians and the Habsburgs, both in 1741, when the
     action is set, and in 1885, when it was first performed.

13   A biography of Barlow, the most gifted and ferocious of the Union Volunteer generals,
     and his remarkable first wife Arabella, remains a curious gap in the extensive literature of
     the Civil War. I am grateful to my wife for the references from her own research notes.

THE REMONSTRANCE REVISITED: ENGLAND AND IRELAND IN THE EARLY FOURTEENTH CENTURY

J.R.S. Phillips

Superficially England and Ireland *circa* 1300 could not have been more different. England was a monarchy whose authority was generally accepted and whose origins went back long before the Norman Conquest of 1066. The kingdom of England was governed by an efficient bureaucracy, with central institutions at Westminster and a network of royal officials in the localities. The operations of this government were meticulously filed away or were recorded on great rolls of parchment, many of which still survive both to fascinate and to torment the twentieth-century historian who attempts to wade through their bulk.[1] Ireland, on the other hand, was, to English eyes, especially to those of English administrators, a land of disorder,[2] to such an extent that the rare moments of peace were to be regarded as a blessing from on high.[3]

Although Ireland had been placed under the authority of the English crown by the papacy during the reign of Henry II, that authority had never been exercised effectively or without question throughout the whole of Ireland.[4] Many of the descendants of Irish rulers who had submitted to Henry II were still powerful and claimed to be kings in their own right.[5] The loyalty of the English colonists could not be taken for granted either and, to make matters worse, some of them were showing conspicuous signs of adopting or adapting to Irish forms of social organisation, speech and dress. Only a few parts of Ireland were firmly under the control of the English royal administration based in Dublin Castle, but even there the mountains which were readily visible from the walls of the Castle harboured Irish enemies, whose depredations were a constant reminder of the fragility of English rule. It was as if the mountains of Wales had overlooked Westminster, and Llywelyn ap Gruffydd or some other Welsh prince had posed a direct threat to the repose of the king of England in his own home.[6]

In reality, however, the contrasts between England and Ireland may have been less sharply defined than contemporary administrators and many twentieth-century academic observers would like to think. When the Normans came to England in 1066 they achieved a thorough-going conquest of the land. By about 1100 no English secular or ecclesiastical lord of any importance still held power, while the work of government was conducted exclusively in either Latin or French. To all appearances England and the English had been submerged under a Norman tide.[7] At the start of the fourteenth century the situation was superficially much the same. The people who mattered in English society still spoke French and government still did its business in Latin or French. None the less, the situation had changed. There is occasional evidence to suggest that the people who mattered, including the king himself, were capable of speaking English when occasion demanded, and may have been doing so as early as the twelfth century.[8] Three successive kings of England, between 1272 and 1377, bore an English rather than a French name, associated with Edward the Confessor, the last but one native king of England before the coming of the Normans. Since his canonisation in 1161 Edward the Confessor had been the patron saint of the rulers of England, whose reputation and legitimacy gained in unquantifiable but significant ways from their holy predecessor, while Westminster Abbey, which Edward himself had rebuilt and in which he was buried in 1066, was reconstructed to house his shrine by Henry III in the thirteenth century.[9] Does this mean that the Norman colonists of England were adopting or adapting to the native culture as the Norman and English colonists of Ireland were apparently doing in the thirteenth century? Ought the historians of medieval England to speak of an 'English revival' in thirteenth-century England to parallel the 'Gaelic revival' at the same period which has been identified by the historians of medieval Ireland? Or is it a matter of perspective and of one's choice of words?[10]

The efficiency of English government and the political unity of the kingdom of England may also to some extent have been illusory. It is easy to be misled by the sheer quantity of surviving government records into assuming that all orders were carried out, all taxes collected, or all crimes punished. Successful government was all too dependent on the personality and energy of the king, the ability of his chief advisers, and the amount and nature of the business to be transacted. It may have been possible to prosecute a war successfully, but in the process the administration of justice and a thousand and one other tasks were likely to be neglected; and the longer a war went on the worse the situation would become.[11] There is some reason to suspect that the kings of medieval England (and perhaps other rulers as well, both secular and ecclesiastical) were taken in by the splendours of their office into believing that their powers of decision and of command were greater than they really were.[12] This may in part explain the troubles at the end of the reign of Edward I, the king who is generally seen as bringing medieval English government to the peak of its efficiency;[13] and it may also help to explain the disorders during the reign of his son Edward II, a man who was in any case unsuited by personality and by inclination for the work of government. Edward II was deposed early in 1327 and probably murdered a few months later,[14] but the crises of his reign were far from unique. The frequency of civil war in thirteenth- and fourteenth-century England, in the reigns of John, Henry III, Edward II and Richard II, not to mention the wars of the fifteenth century, and of the royal depositions which often followed, tell some kind of story, and it may be that it is not one of steadily developing political liberties and orderly government, as traditional constitutional historians used to argue, but rather that of a society which was inherently unstable and prone to outbursts of extreme violence among the members of its ruling class. Was there really then so big a difference between the kingdom of England and the lordship of Ireland? Should we perhaps see the ambitions of a Donal O'Neill or a rebellious Earl of Desmond within Ireland in a similar light to those of a Simon de Montfort or a Thomas of Lancaster within England? The conventional division of Ireland into 'the land of peace' and 'the land of war' may perhaps have been a more accurate and honest description of political reality[15] than the united kingdom of England, where the king's peace supposedly reigned supreme, but where local disturbances were endemic and the king himself could be the ultimate victim.[16] It has been aptly remarked of England that even in the best ordered of reigns the forces of authority were barely adequate and 'the internal peace of the kingdom was poised on a razor's edge'.[17] These comparisons between England and Ireland are oversimplified, not to say simplistic, but they do raise serious issues and serve as a warning that what we see in a society is partly dependent on what we expect to see and on the conventional language that we use to describe it.

One important difference between thirteenth-century England and Ireland is in the respective implications of an 'English revival' and a 'Gaelic revival'. In the case of England the re-emergence of English as a spoken and a literary language betokened a major cultural change within the ruling classes of the kingdom; but it did not carry with it any possibility of the restoration of a native English political power. There was no descendant of the house of Wessex, or of an earl of Mercia, waiting in the wings to press his claims, for all such power had been destroyed by the Norman Conquest. In Ireland on the other hand a 'Gaelic revival' could have political implications in, for example, the inauguration of Donal Mac Murrough as king of Leinster in

1327, or in the attempts to revive the High Kingship of Ireland on behalf of Brian O'Neill in 1258–60, King Haakon of Norway in 1263 and, finally, Edward Bruce of Scotland in 1315.[18]

Edward Bruce's inauguration as king of Ireland in June 1315 followed swiftly on his landing in the English-ruled Lordship of Ireland at the end of May that year.[19] The timing of the Scottish invasion of Ireland was probably influenced to some extent by a desire to press home the advantage gained by Robert Bruce's defeat of Edward II of England at Bannockburn in June 1314. A 'second front' in Ireland might, as Professor Lydon has suggested, force the English to divert military and financial resources from England to Ireland.[20] Although in the end England sent little or nothing to Ireland, because there was little or nothing available to send, there is no doubt that English garrisons in the North of England suffered miseries through the absence of supplies of Irish food and money which had made a large contribution to the English war effort since the start of the conflict with Scotland in 1296.[21] However, the Scottish invasion of Ireland was also the result of long-standing connections between Scotland and Gaelic Irish society.[22] The immediate occasion of Edward Bruce's coming to Ireland may well have been some kind of spontaneous or, more likely, engineered invitation from Donal O'Neill, king of Tyrone, the son of the unsuccessful candidate for the high kingship in 1258;[23] but it is quite likely, as has recently been suggested, that Scottish ambitions in Ireland can be traced back well before 1315, perhaps as far as 1306–7 at the time of Robert Bruce's brief exile from Scotland on Rathlin island.[24]

The three years between Edward Bruce's landing in Ireland in May 1315 and his death at Faughart in October 1318 were marked by war, devastation and famine.[25] But they were also marked by a resolute attempt on the part of Edward Bruce's chief Irish supporter, Donal O'Neill, to employ the technique of written propaganda to persuade Pope John XXII in late 1317 to sanction the transfer of authority over Ireland from the English crown to Edward Bruce of Scotland. The resulting document, commonly though inaccurately known as the Remonstrance of the Irish Princes,[26] was in effect a detailed commentary on the alleged failure of the English crown and its officials, the English clergy, and the English settlers in Ireland to observe the terms of the papal bull *Laudabiliter* of about 1155 in which Pope Adrian IV had authorised Henry II of England to come to Ireland 'to enlarge the boundaries of the church, to reveal the truth of the Christian faith to unlearned and savage peoples, and to root out from the lord's field the vices which grow in it.'[27]

In a famous article in 1957 Professor J.A. Watt noted that the Remonstrance was not the only fourteenth-century occasion on which an appeal was made to the authority of *Laudabiliter*.[28] It was also referred to in the list of grievances against English rule which was lodged by the Irish Church at the Council of Vienne in 1311;[29] and in 1331 the Justiciar of Ireland cited *Laudabiliter* when he wrote to the pope complaining of the alleged ill-treatment of the English by the Irish.[30] But *Laudabiliter* was also made use of on other occasions. At some point in the late thirteenth or early fourteenth century an unknown Irish scribe entered a copy of *Laudabiliter* into the twelfth-century Irish manuscript known as the Book of Leinster, prefacing his transcript with the very significant remark that nothing in the bull in any way diminished the rights of the Irish;[31] in about 1318 a group of the English colonists in Ireland sent a copy of *Laudabiliter* to Edward II and his council in England, together with a petition in which they complained about the behaviour of the Irish towards the English;[32] and in 1347 one of the charges against Maurice Fitz Thomas, the famous 'Rebellious Earl of Desmond', was that he had informed the pope that Edward III of England was not observing the terms of *Laudabiliter* and had implied that Edward should therefore be deprived of his authority over Ireland.[33] The number of references to *Laudabiliter* shows that both the English and the Irish population of Ireland were aware of the propaganda value that could be extracted from the bull, while the concentration of these references between about 1311 and 1331 suggests that an Irish appeal to *Laudabiliter* tended to provoke an English counter-appeal and vice versa.[34]

A second important point about the Remonstrance is the question of who composed it. This may seem a futile exercise since there is no hint in the document itself, other than the opening statement that it was initiated by Donal O'Neill and the princes of Ireland, as to its author, and there is no surrounding body of documentary evidence to provide a context.[35] Nonetheless there are some clues, at least to the kind of man who might have written the Remonstrance. An analysis of the Latin text shows considerable rhetorical skill and that the author made extensive use of the cursus, the special metrical prose of the Roman Curia,[36] probably in the hope that, if the pope issued any letters in response to the Remonstrance, he would incorporate some of its language. This strongly suggests a well educated clerical author, although it is not of course absolute proof. A further clue is that the author's descriptions of the atrocities committed by the English secular lords in Ireland do not refer to very recent events, the latest having occurred in 1305.[37] By contrast the full power of his invective was reserved for the contemporary behaviour of the English clergy in Ireland: for Walter Jorz, an English Dominican, who had been Archbishop of Armagh between 1307 and 1311, whom he described as 'a man of small wit and no learning';[38] for the Cistercian monks of Abbeylara and of Inch whom he accused of celebrating masses after attacking and slaying the Irish;[39] and for a certain Franciscan friar named Simon who allegedly claimed in the presence of Edward Bruce himself that it was no sin to kill an Irishman.[40] Friar Simon was probably Simon le Mercer of the Franciscan house at Drogheda who made two visits to Edward II in England in 1317, and it is possible that he had also visited Edward Bruce as an official or semi-official envoy on behalf of the governments in Dublin and in England.[41] This evidence, such as it is, suggests that the author of the Remonstrance was a cleric, that he had a particular interest in the church of Armagh, and that he had a special hostility towards religious orders like the Cistercians and Franciscans which contained a large number of English members.[42]

There matters would rest, but for a stroke of luck. In 1303 a strange set of events began, following the death on 10 May of the powerful ecclesiastical figure of Nicholas Mac Mael Íosa, archbishop of Armagh since 1272.[43] On about 31 August 1303 a certain Michael Mac Lochlainn, who held the office of 'lector' or 'reader' at the Franciscan house in Armagh, was elected as archbishop by the dean and chapter of Armagh. In October 1303 Edward I gave royal assent to the election and authorised Mac Lochlainn to go to the papal Curia to seek confirmation of his appointment.[44] However the newly elected pope, Benedict XI, refused to do so, ostensibly on the grounds of his illegitimacy,[45] but probably also because Franciscans, whether Irish or not, were in bad odour at the Curia on account of the controversy over the Spiritual Franciscans which was then coming to its climax. At the time of Benedict's own election at Perugia in October 1303 a number of Spirituals of a Messianic turn of mind were hopefully prophesying the election of an 'angelic pope' in the mould of Celestine V to succeed Boniface VIII, while one of their number, Ubertino da Casale, who had described Boniface as 'the mystical Antichrist', publicly identified Benedict XI as 'one of the beasts of the Apocalypse'.[46] In Mac Lochlainn's place the pope provided Denis, the dean of Armagh, who may have been one of those who accompanied Mac Lochlainn to the curia. For unknown reasons Denis was not consecrated as archbishop and resigned the office in about 1304. The see then remained vacant until the appointment by Clement V in August 1306 of John Taaffe who died about a year later.[47] On 6 August 1307 Clement V appointed the Dominican, Walter Jorz, the archbishop who was the object of attack in the Remonstrance and who remained in office until his resignation in 1311.[48] His successor was his own brother, Roland Jorz, also a Dominican, who was provided and consecrated as archbishop on 13 November 1311.[49] Roland Jorz's career as archbishop is something of a mystery since he may have been suspended from office in 1316.

He finally resigned in 1322 and ended his days as a suffragan bishop in the province of Canterbury.[50]

The general significance of all these events was that the death of Nicholas Mac Mael Íosa in 1303 marked the end of the tenure of the see of Armagh in the medieval period by archbishops of native Irish birth.[51] With the appointment of Taaffe in 1306 and, more especially, of Walter Jorz in 1307 Armagh fell under English control. Walter Jorz was a former royal clerk, while Walter's brother, Thomas, who was also a Dominican, had been chaplain to Edward I before he was made a cardinal by Clement V in December 1305.[52] Under such circumstances an Irish clerk had little chance of succession to Armagh. The particular significance may have more than a little to do with the Armagh Franciscan, Michael Mac Lochlainn, who was clearly a man of ambition who resented being passed over for the archbishopric of Armagh, all the more so perhaps because the pope who rejected him and the brothers, Walter and Roland, who eventually enjoyed the office were all Dominicans. In 1310 Mac Lochlainn was given permission by Clement V, despite his illegitimacy, to seek any ecclesiastical office, including that of archbishop of Armagh,[53] and finally gained preferment in 1319 when he was appointed as bishop of Derry, where he remained until his death in 1349.[54]

Michael Mac Lochlainn was exactly the sort of person who could have composed the Remonstrance: a cleric with an academic training;[55] and a man whose ambitions for high office had been disappointed, but who still hoped for advancement. He was also a member of a religious order which racially was very deeply divided[56] and whose Irish members were strongly suspected by the English government in Ireland of preaching in support of the Scots during the Bruce invasion.[57] It is also possible that Mac Lochlainn belonged to one of those ecclesiastical dynasties which were a feature of the medieval Irish church, since earlier in the fourteenth century Gofraid Mac Lochlainn, who was presumably a relation, had been bishop of Derry.[58] His secular relationships are more speculative, but he may have been a member of the family of Mac Lochlainn, one of whom, Muirchertach Mac Lochlainn, had been High-King between 1145 and 1166, and who had held the kingship of Cinéal Eóghain until 1241. Their successors in Cinéal Eóghain were a related family, the O'Neills, of which Donal O'Neill, the man nominally responsible for the production of the Remonstrance, was the latest representative.[59]

Assuming that Michael Mac Lochlainn really was the author of the Remonstrance, which is a plausible but probably unprovable hypothesis, we would have a man who was connected in very significant ways with both Irish political and ecclesiastical society at a very important phase of Irish history, the Bruce invasion. But Mac Lochlainn, if it were he, perhaps did more than just compose a document justifying the kingship of Edward Bruce for onward transmission to the pope. It is possible that he already had some kind of dossier of English secular and ecclesiastical offences against the Irish, which he had been collecting since his disappointment at not becoming archbishop of Armagh in 1303. The summoning of the general council of the Church to meet at Vienne in 1311, when churchmen from all over Western Christendom presented their grievances against their local secular authorities, could also have acted as a stimulus for such a document.[60] Mac Lochlainn may also have seen the Scottish invasion of Ireland as an opportunity to restore the independence of the church of Armagh within the framework of the restored high kingship in the hands of Edward Bruce, and preferably with himself as archbishop. It may be very significant in this connection that the register of the early fifteenth-century Archbishop Fleming contains a transcript of a document dated November 1315 in which Donal O'Neill and his son granted privileges to the church of Armagh which were to be guaranteed by Edward Bruce, King of Ireland.[61]

Whether the Remonstrance was simply a propaganda device to persuade the pope to approve a change of political regime in Ireland or whether it was also, and perhaps more profoundly, the product of a deeply and cunningly laid ecclesiastical plan mattered little in the short term. Michael Mac Lochlainn lived to enjoy his diocese of Derry but in October 1318 Edward Bruce and many of his followers were defeated and killed by an army of the English colonists at Faughart near Dundalk. With them died the dream of reshaping the political map of medieval Britain and Ireland. The English-ruled lordship of Ireland, which Bruce and O'Neill had sought to destroy, survived until 1541 when it was finally extinguished at the hands of the king of England himself.[62]

NOTES

1   For the most recent discussions of medieval English government, see W.L. Warren, *The Governance of Norman and Angevin England, 1086–1272* (London 1987) and A.L. Brown, *The Governance of Late Medieval England, 1272–1461* (London 1989).

2   See, for example, the complaints about the state of Ireland made by the Justiciar Geoffrey de Geneville in the mid 1270s and those of the Justiciar John Morice, in the early 1340s: London, PRO, S.C.1/5/96; ibid./18/12, 13, 16; ibid./38/108.

3   In a postscript to a letter written early in the reign of Edward I the unidentified author delivered himself of the following cry from the heart: 'Benedictus auctor pacis altissimus pacificata est terra Hibernie adeo quod in singulis locis eiusdem vigent pax et tranquillitas hiis diebus', ibid./62/41.

4   See R. Frame, *The Political Development of the British Isles, 1100–1400* (Oxford 1990), pp. 85–9; R.R. Davies, 'Lordship or Colony?', in J.F. Lydon (ed.), *The English in Medieval Ireland* (Dublin 1984).

5   See, for example, J.F. Lydon, 'Lordship and Crown: Llywelyn of Wales and O'Connor of Connacht', in R.R. Davies (ed.), *The British Isles, 1100–1500: Comparisons, Contrasts and Connections* (Edinburgh & Atlantic Highlands, NJ 1988), especially p. 60; R. Frame, *Political Development of the British Isles*, pp. 108–15; R.R. Davies, *Domination and Conquest: the Experience of Ireland, Scotland and Wales, 1100–1300* (Cambridge 1990), especially ch. 3, 'Native Submission'.

6   On the general political situation in Ireland in the late thirteenth and early fourteenth centuries, see the chapters by J.F. Lydon, 'The years of crisis, 1254–1315', and 'A land of war', in Art Cosgrove (ed.), *A New History of Ireland* (henceforward *NHI*), ii, *Medieval Ireland, 1169–1534* (Oxford 1987), pp. 179–204, 240–74.

7   See R. Frame, *Political Development of the British Isles*, ch. 1, 'The British Isles in 1100: Political Perceptions and the Geography of Power'; M. Chibnall, *Anglo-Norman England, 1066–1166* (Oxford 1986); M.T. Clanchy, *England and its Rulers, 1066–1272: Foreign Lordship and National Identity* (London 1983).

8   See, for example, J.R.S. Phillips & E.L.G. Stones, 'English in the Public Records: Three Late Thirteenth-Century Examples', *Nottingham Medieval Studies*, vol. XXXII (1988), pp. 197–206.

9   Frank Barlow (ed.), *The Life of King Edward Who Rests at Westminster* (Oxford 1992) (2nd rev. edn of *The Life of King Edward the Confessor* [London 1962]); especially Appendix D, 'The Development of the Cult of King Edward' ; M. Prestwich, *Edward I* (London 1988), pp. 4–5; Clanchy, *England and its Rulers*.

10 The paradox is I think largely a matter of perspective. It would be surprising in both England and Ireland if the conquerors and settlers, who formed a small minority of the population, should not have taken on some of the characteristics of the majority among whom they lived.

11 See, for example, M. Prestwich, *War, Politics and Finance under Edward I* (London 1970), especially ch. XII, 'Politics and the King, 1298–1307', and ch. XIV, 'The Social Consequences of War'; *idem, The Three Edwards: War and State in England, 1272–1377* (London 1980); J.R. Maddicott, *The English peasantry and the Demands of the Crown, 1294–1341, Past and Present*, Supplement 1 (Oxford 1975); *idem*, 'Poems of Social Protest in Early Fourteenth-Century England', in W.M. Ormrod (ed.), *England in the Fourteenth Century: Proceedings of the 1985 Harlaxton Symposium* (Woodbridge 1986); W.M. Ormrod, 'The Crown and the English Economy, 1290–1348', in Bruce M.S. Campbell (ed.), *Before the Black Death: Studies in the 'Crisis' of the Early Fourteenth Century* (Manchester 1991); J. Taylor and W. Childs (eds), *Politics and Crisis in Fourteenth-Century England* (Gloucester 1990).

    For a stimulating attempt to question the effectiveness of English government at an earlier period, see W.L. Warren, 'The myth of Norman administrative efficiency', *Transactions of the Royal Historical Society (TRHS)*, 5th series, 34 (1984), pp. 113–32.

12 The classic example in the late thirteenth century was the pope Boniface VIII (1294–1303), who inherited a complex administrative machine and system of canon law which were designed to emphasise and to perpetuate the claims of the papacy to universal authority. His famous conflicts with Philip IV of France (1285–1314) and Edward I of England (1272–1307) showed how illusory these claims were in practice. Edward I himself found his ambitions checked by his prolonged wars with France and Scotland and by

political troubles at home. Philip IV's successes generated tensions and were swiftly followed after his death in 1314 by political turmoil within France.

13   For the most recent discussion of the administrative achievements of Edward I and his ministers, which were real enough, see Prestwich, *Edward I*, especially ch. 9, 'The Government of England, 1278–86', and ch. 10, 'The Statutes and the Law'; the multitude of problems which Edward faced in the closing years of his reign can be studied in ibid., ch. 19, 'The Last Years, 1298–1307' as well as in *idem, War, Politics and Finance*.

14   This is a theme which I hope to treat in more detail elsewhere. For the complex politics of this period, see J.R.S. Phillips, *Aymer de Valence Earl of Pembroke, 1307–1324: Baronial Politics in the Reign of Edward II* (Oxford 1972); J.R. Maddicott, *Thomas of Lancaster, 1307–1322* (Oxford 1970); N. Fryde, *The Tyranny and Fall of Edward II, 1321–1326* (Cambridge 1979). The unsuitability of Edward II for his role as king is shown very clearly in the episode between 1317 and 1319 of the Holy Oil of St Thomas of Canterbury which he hoped would bring about a miraculous change in his political fortunes; equally revealing was the appearance at Oxford in July 1318 of John Powderham, an imposter claiming to be the real king. See J.R.S. Phillips, 'Edward II and the Prophets' in Ormrod, *England in the Fourteenth Century*, pp. 196–201; W. Childs, '"Welcome, my brother": Edward II, John of Powderham and the Chronicles, 1318', in I. Wood & G.A. Loud (eds), *Church and Chronicle in the Middle Ages: Essays Presented to John Taylor* (London 1991), pp. 149–63.

15   See Robin Frame's important paper, 'War and Peace in the Medieval Lordship of Ireland', in J.F. Lydon (ed.), *The English in Medieval Ireland* (Dublin 1984), pp. 118–41.

16   The classic case of a local breakdown of law and order is that of the Folville family from Leicestershire whose criminal career ran unchecked for over twenty years during the reign of Edward III, a 'strong' king, see E.L.G. Stones, 'The Folvilles of Ashby-Folville and their Associates in Crime, 1326–47', *TRHS*, 5th series, vol. 7 (1957); J.G. Bellamy, 'The Coterel gang', *English Historical Review*, vol. 79 (1964). During the reign of Edward II there are many examples such as the rebellion by the city of Bristol against royal taxation which was ended only by a siege in 1316; the attack in 1318 on the earl of Pembroke's manor of Painswick in Gloucestershire by members of the locally powerful Berkeley family, his former retainers; the ravages of Gilbert de Middleton in Northumberland and Durham in 1317–18, which included the robbing of two papal envoys; and the earl of Lancaster's attacks on the earl of Surrey's Yorkshire lands at Conisborough and Sandale and his Welsh marcher lordships of Bromfield and Yale in 1318: Phillips, *Aymer de Valence*, pp. 102–3, 261–6, 125–8, 165, 171; Maddicott, *Thomas of Lancaster*, pp. 204–8, 234–6.

17   Phillips, *Aymer de Valence*, p. 124, citing R.L. Storey, *The End of the House of Lancaster* (London 1966), p. 21.

18   *NHI*, ii, pp. 302, 244–6, 248; E. Curtis, *A History of Medieval Ireland* (London 1938), pp. 144–5, 149.

19   A.A.M. Duncan has argued persuasively for this redating of Edward Bruce's inauguration in his paper, 'The Scots' Invasion of Ireland, 1315', in Davies, *The British Isles*, pp. 109–10.

20   J.F. Lydon, 'The Bruce invasion of Ireland', *Historical Studies*, IV (1963), pp. 112–3; Professor Frame is however sceptical of this since the Scots would themselves face problems of supply once they reached Ireland: R. Frame, 'The Bruces in Ireland, 1315–18', *Irish Historical Studies (IHS)*, xix, no.73 (March 1974), pp. 9–10. My own reservations about this explanation are indicated in the following note.

21   See J.R.S. Phillips, 'The Mission of John de Hothum to Ireland, 1315–1316', in J.F. Lydon (ed.), *England and Ireland in the Later Middle Ages: Essays in Honour of Jocelyn Otway-Ruthven*

(Dublin 1981), pp. 62–3, 74–6, and J.F. Lydon, 'Edward I, Ireland and the War in Scotland, 1303–1304', in ibid.; J.R.S. Phillips, 'Documents on the Early Stages of the Bruce Invasion of Ireland, 1315–1316', *Proceedings of the Royal Irish Academy (RIA Proc.)*, 79 C (Dublin 1979), pp. 250–1, 265–6.

22 S. Duffy, 'The Bruce Brothers and the Irish Sea World, 1306–29', *Cambridge Medieval Celtic Studies*, no. 21, summer 1991, pp. 59–70; J.R.S. Phillips, 'The Anglo-Norman Nobility', in Lydon, *The English in Medieval Ireland*, p. 102.

23 The exact circumstances of the Scottish invasion of Ireland will probably never be known. While the Irish tract *Cath Fhocairte Brighte*, which refers to an invitation from Ireland, has now been shown to be a nineteenth-century fabrication, this does not rule out such an invitation. The fact that Edward Bruce was inaugurated as king of Ireland so soon after his arrival is in itself a strong indication of a previously laid plan: see S. Duffy, 'The Gaelic account of the Bruce invasion *Cath Fhochairte Brighite*: medieval romance or modern forgery?', *Seanchas Ard Mhacha*, vol. 13, no. 1 (Dundalk 1988), and J.R.S. Phillips, 'The Irish Remonstrance of 1317: an International Perspective', *IHS*, xxvii, no. 106 (November 1990), p. 126 n. 57; S. Duffy, 'The "Continuation" of Nicholas Trevet: a New Source for the Bruce Invasion', *RIA Proc.*, 91 C. no. 12 (Dublin 1991), pp. 308–9, 314.

24 Duffy, 'The Bruce Brothers', pp. 64–5, 70–6; see also Lydon, 'The Bruce Invasion', pp. 113–15.

25 The best accounts of the invasion are to be found in Lydon, ibid.; Frame, 'The Bruces in Ireland', pp. 3–37; S. Duffy, 'The Bruce Brothers'; and Duncan, 'The Scots' Invasion of Ireland, 1315'.

26 For a recent attempt to put the Remonstrance into the wider context of the relations between England, Scotland, Wales and the papacy in the late thirteenth and early fourteenth centuries, see J.R.S. Phillips, 'The Irish Remonstrance of 1317', pp. 112–29. As is well known, the Remonstrance is preserved only in fourteenth- and fifteenth-century Scottish chronicles, the *Chronica Gentis Scotorum* of John Fordun and the *Scotichronicon* of Walter Bower. Hitherto the only available translation of the Remonstrance was that in E. Curtis & R.B. McDowell (eds), *Irish Historical Documents 1172–1922* (London 1943), pp. 38–46, based on a text in the 1722 edition of Fordun by Thomas Hearne. However there is good reason to believe that the text preserved by Walter Bower is, although later in date, derived from a better manuscript tradition. This version of the Remonstrance, edited by Norman F. Shead and newly translated by the late Martin S. Smith, together with a commentary by J.R.S. Phillips, can be found in D.E.R. Watt (ed.), *Scotichronicon by Walter Bower*, volume 6, Books XI and XII (Aberdeen 1991), pp. xxi–xxv (Introduction), 384–403 (text and translation), 465–81 (commentary). The Remonstrance is followed by a text of *Laudabiliter*: ibid., pp. 403–4 (text and translation), 481–3 (commentary).

27 Ibid., p. 403. In very powerful language the Remonstrance listed some of the atrocities committed by the English against the Irish, such as the murders of Art and Muirchertach Mac Murrough by Geoffrey de Pencoyt in 1282 and of the O'Connors of Offaly by Piers de Bermingham of Tethmoy in 1305: ibid., pp. 394–5. For a detailed examination of these and other examples cited in the Remonstrance, see the commentary in ibid., pp. 465–81.

28 J.A. Watt, '*Laudabiliter* in medieval diplomacy and propaganda' in *Irish Ecclesiastical Record*, 5th ser., lxxxvii (January-June 1957), pp. 420–32.

Strictly speaking *Laudabiliter* had no direct bearing on the existence of the English-ruled Lordship of Ireland since Henry II's campaign in Ireland in 1171–2 occurred under very different circumstances, while the Lordship itself did not come into being until 1185 when Henry II bestowed it upon his youngest son John. None the less *Laudabiliter* and the Lordship came to be regarded as part and parcel of one another.

29   J.A. Watt, *The Church and the Two Nations in Medieval Ireland* (Cambridge 1970), pp. 143–4, citing the complaints of the Irish clergy (and those of other provinces of the Western Church) published in 'Ein Bruchstück der Akten des Konzils von Vienne', P. Ehrle (ed.), *Archiv für Literatur und Kirchengeschichte*, iv (1888), p. 370.

30   Watt, pp. 196–7; and *idem*, 'Negotiations Between Edward II and John XXII Concerning Ireland', *IHS*, x (1956), pp. 18–20.

31   R. Atkinson (ed.), *The Book of Leinster* (Dublin 1880), p. 342: this is a facsimile of the original. The hand in which *Laudabiliter* is written is consistent either with the early fourteenth-century date suggested by W. O'Sullivan in notes on the scripts and make-up of the *Book of Leinster*, *Celtica*, vii (1966), p. 3, or with a slightly (but not much) earlier date.

32   This version is preserved in PRO SC8/177/8818. The text is very similar, though not identical, to the text of *Laudabiliter* copied into the *Book of Leinster* and to the text which accompanied the Remonstrance to Avignon in 1317–18. There are however sufficient differences between these three texts and the two almost identical ones contained in the twelfth-century *Expugnatio Hibernica* of Gerald of Wales and the *Ymagines Historiarum* of Ralph of Diceto to suggest that there were at least two separate modes of transmission of *Laudabiliter*: see Bower, *Scotichronicon*, pp. 403–4, 481–2.

        The petition which accompanied the PRO text of *Laudabiliter* was sent to the English crown by *Lestat Dirlande* (evidently representing the colonial population and not the native Irish) at a date between about 1317 and 1319, and therefore close to the time of composition of the Remonstrance itself. The petition (SC8/177/8820), is printed in G.O. Sayles (ed.), *Documents on the Affairs of Ireland Before the King's Council* (Dublin 1979; recte 1980), pp. 99–101. In close proximity to the petition and *Laudabiliter* are two other documents, a petition from the mayor and citizens of Dublin concerning their arrest of the earl of Ulster, and a petition from the Justice of Wales about the dangers of attack by Scottish allies of the Irish on the harbours of Anglesey (SC 8/177/8817, 8819). It is clear from the hand and from the contents that all four documents once formed a group on their own relating to Irish matters *c.*1317–18.

33   G.O. Sayles, 'The Legal Proceedings Against the First Earl of Desmond', *Analecta Hibernica*, no. 23 (Dublin 1966), pp. 20, 44. I should like to thank Professor Robin Frame for drawing my attention to this use of *Laudabiliter*.

34   Cf. Frame, *Political Development of the British Isles*, p. 187.

35   The suggestion that the Remonstrance might have been written by the same person as the Scottish Declaration of Arbroath of 1320 was never very plausible: see R.G. Nicholson, 'Magna Carta and the Declaration of Arbroath', *Edinburgh University Journal*, xxii (1965), p. 143; Phillips, 'The Irish Remonstrance of 1317', pp. 124, 127. Although it will be argued here that the Remonstrance was composed by an Irish author, it should be remembered that the primary purpose of the document was to obtain papal sanction for Edward Bruce's inauguration as king of Ireland. This implies some Scottish involvement in the composition or at least approval of its contents and line of argument: see Bower, *Scotichronicon*, p. 480. On the lack of a documentary context, see Phillips, 'The Irish Remonstrance of 1317', pp. 118–19.

36   In an English translation the Remonstrance appears as a very untidy and rambling document, passages of anti-English rhetoric being interspersed with circumstantial charges like the ones mentioned above (n. 27). While I am still of the opinion that the Remon-

strance in its final form was probably composed in a hurry (ibid., pp. 125–7), a closer examination of the Latin text suggests that the author was highly skilled. My colleague Dr Michael Haren of the Irish Manuscripts Commission has also shown that the Remonstrance contains at least thirty examples of the *Cursus Velox*, about twenty of the *Cursus Planus*, and a few of the *Cursus Tardus*: see Bower, *Scotichronicon*, p. xxii.

37   They all relate to events which had occurred in Leinster or in the Midlands of Ireland well before the composition of the Remonstrance, whereas the ecclesiastical incidents all took place within the province of Armagh, in the vicinity of the areas most directly affected by the Bruce invasion: see Bower, *Scotichronicon*, pp. 392–5, 474–6.

38   Ibid., pp. 392–3, 472–3; M.H. McInerney, O.P., *A History of the Irish Dominicans* (Dublin 1916), pp. 507–603.

39   Bower, *Scotichronicon*, pp. 396–7, 476. Abbeylaragh (also called Granard or Laragh), Ardagh diocese, province of Armagh, was founded *c*.1210 by Sir Richard Tuit and colonised by monks from St Mary's abbey, Dublin, in 1214. Inch (also called Iniscourcey), diocese of Down, province of Armagh, was founded by John de Courcy in 1180 or 1188, and filled with monks from Furness in England (A. Gwynn & R.N. Hadcock, *Medieval Religious Houses* [London 1970], pp. 121–2, 124, 135).

40   Bower, *Scotichronicon*, pp. 396–7, 476–7.

41   Ibid.; London, Society of Antiquaries Ms.120/f.10; ibid. Ms.121/f.2 (Wardrobe Books of 10 & 11 Edward II); E.B. Fitzmaurice & A.G. Little (eds), *Materials for the History of the Franciscan Province of Ireland A.D. 1230–1450* (Manchester 1920), pp. 101–2. Friar Simon may have played a role during the period of the Bruce invasion similar to that of the better known Philip of Slane, O.P. (Frame, 'The Bruces in Ireland', p. 34, n. 144A).

42   It may be more than coincidence that three of the Anglo-Norman lords who were most bitterly attacked by the author of the Remonstrance, Thomas de Clare (d.1287), Piers de Bermingham (d.1308), and John Fitz Thomas, Earl of Kildare (d.1316), were buried in Franciscan houses of English allegiance (i.e. Limerick and Kildare) (Fitzmaurice & Little, *Materials*, pp. 58–9, 88–9, 101).

43   On his career, see A. Gwynn, 'Nicholas Mac Maol Íosa, archbishop of Armagh (1272–1303)', in J. Ryan (ed.), *Essays and Studies Presented to Professor Eoin MacNeill* (Dublin 1940), pp. 394–405.

44   Fitzmaurice & Little, *Materials*, p. 80.

45   Ibid. Mac Lochlainn was allegedly the son of a nun (*Calendar of Papal Registers: Papal Letters*, ii *[1305–1342]* [London 1895], p. 72). In the light of later events surrounding the church of Armagh it is significant that Edward I appears to have had no objection to another Irishman succeeding Nicholas as archbishop.

46   M.E. Reeves, 'Some popular prophecies from the fourteenth to the seventeenth centuries', in G.J. Cuming & D. Baker (eds), *Popular Belief and Practice, Studies in Church History* (Cambridge 1972), viii, pp. 107–8; M.D. Lambert, *Franciscan Poverty: the Doctrine of the Absolute Poverty of Christ and the Apostles in the Franciscan Order, 1210–1323* (London 1961), pp. 174–6; D.L. Douie, *The Nature and the Effect of the Heresy of the Fraticelli* (Manchester 1932), pp. 138–40.

47   Fitzmaurice & Little, *Materials*, p. 80; See T.W. Moody, F.X. Martin and F.J. Byrne (eds), *A New History of Ireland*, ix, *Maps Genealogies, Lists* (Oxford 1984) p. 269.

48   Ibid. Clement V informed Edward II of the appointment and consecration of Walter Jorz in a bull dated from Poitiers on 6 August 1307 (P.N.R. Zutshi, *Original Papal Letters in England, 1305–1415; Index Actorum Romanorum Pontificum ab Innocentio III ad Martinum V*

Electum, CISH – Commission Internationale de Diplomatique, Biblioteca Apostolica Vaticana, [Vatican City 1990], p. 24, no. 47; the original bull is preserved as PRO, SC7/44/22).

49   *NHI*, ix, p. 269; Zutshi, p. 33, no. 67; the original bull is PRO, SC7/44/15. The bull announcing the appointment and consecration of Roland Jorz was dated at Vienne during the time of the general council of the Church that was being held there. Possibly the resignation of Walter and his replacement by Roland had something to do with the business of the council in relation to Ireland.

50   *NHI*, ix, p. 269. He was however present with the leaders of the Anglo-Irish army of John de Bermingham at the battle of Faughart in 1318 (S. Duffy, 'The "Continuation" of Nicholas Trevet', p. 314).

51   Gwynn, 'Nicholas Mac Maol Íosa', p. 394.

52   MacInerney, *History of the Irish Dominicans*, p. 507; J.H. Denton, *Robert Winchelsey and the Crown, 1294–1313: a Study in the Defence of Ecclesiastical Liberty* (Cambridge 1980), p. 233.

53   *Cal. Papal Registers*, II (1305–42), p. 72. Presumably Mac Lochlainn had first of all petitioned the pope for such authorisation.

It would be tempting, but altogether too neat, to identify Michael Mac Lochlainn with the 'Brother Malachy of Ireland' who, according to Luke Wadding, preached before Edward II and his court in that same year and excoriated them for their sins (Wadding, *Annales Minorum*, vol. 6 [Rome 1733], p. 176). Unlike Malachy, who is said to have studied theology at Oxford (ibid.), there is no evidence that Michael Mac Lochlainn was an Oxford graduate. (see also n. 55 below). It is possible that Malachy should be identified with the famous Malachy of Limerick, the unsuccessful candidate for the archdiocese of Tuam in 1286 and the probable author of the treatise *De Veneno* on the seven deadly sins: Fitzmaurice & Little, *Materials*, pp. 54–8.

54   He was elected after 19 August 1319 and died about 18 December 1349 (*NHI*, ix, p. 279).

55   His position as *lector* or 'reader' at Armagh clearly implies an academic training, which he could well have received in Paris. The General Constitutions of 1260 made provision for two Irish members of the order to study at Paris without requiring special permission to do so (Fitzmaurice & Little, *Materials*, p. 79; Watt, *The Church and the Two Nations*, p. 180).

Other examples of Franciscans who held the office of *lector* are the philosopher Duns Scotus at Cologne in 1308 and the missionary traveller William of Rubruck at Acre in the kingdom of Jerusalem in the 1250s and possibly at Nicosia in Cyprus before that (Fitzmaurice & Little, *Materials*, p. 87; *The Mission of Friar William of Rubruck: his Journey to the Court of the Great Khan Mongke, 1253–1255*, P. Jackson and D. Morgan (eds), Hakluyt Society Second Series, no. 173 [London 1990], p. 41).

56   In 1291, for example, there had been sixteen deaths when English and Irish Franciscans came to blows at the general chapter held at Cork (A.J. Otway-Ruthven, *A History of Medieval Ireland* [London 1968], p. 138; Fitzmaurice & Little, *Materials*, pp. 63–4). On this incident and on the general level of hostility within the order, see Watt, *The Church and the Two Nations*, pp. 180–8, 190–3.

57   Signs of distrust of the native Irish Franciscans can be found as far back as the 1280s (ibid., pp. 181–2; Fitzmaurice & Little, *Materials*, pp. 52–3). There was particular uneasiness after 1315. Edward II was unsuccessful in obtaining confirmation of the disputed election of the English Franciscan, Geoffrey of Aylsham, as archbishop of Cashel, but in August 1316 he sent him and the minister of the Irish province to complain to Michael of Cesena, the minister-general of the order, about the sympathy of Irish Franciscans for

Edward Bruce (PRO, C70/3, m. 7; Watt, *The Church and the Two Nations*, pp. 184–5; Fitz-maurice & Little, *Materials*, pp. 94–100 [the extract from the Remonstrance on pp. 95–6 is wrongly dated as *c*.1315–16]).

There was also trouble within the Dominicans in Ireland. In 1318, the general chapter held at Lyons ordered the expulsion from the English province of Brother Henry Glam, *Ibernicum*, for encouraging other friars to resist the authority of their English superiors in Ireland (*Acta Capitulorum Generalium Ordinis Praedicatorum*, vol. 2, [1304–78]; B.M. Reichert (ed.), *Monumenta Ordinis Fratrum Praedicatorum Historica* [Rome 1899], vol. 4, pp. 112–13).

58   Gofraid was bishop of Derry between 1297 and *c*.1315. His successor was Áed Ó Néill, 1316 to June 1319: *NHI*, ix, p. 279. The diocese of Derry had been held between 1185 and 1293 by members of the Ó Cerbhalláin family. This kind of family control was less fre-quent by the end of the thirteenth century than it had been earlier, but was again becom-ing common in the fifteenth century (Watt, *The Church and the Two Nations*, pp. 150–1). For the fifteenth-century examples of the families of Mac Cawell and Maguire in the diocese of Clogher and of Mac Egan in the diocese of Clonfert, see Watt, *The Church in Medieval Ireland* (Dublin 1972), pp. 185–92.

59   K. Nicholls, *Gaelic and Gaelicised Ireland in the Middle Ages* (Dublin 1972), pp. 127–8. If Michael Mac Lochlainn was really connected with the ruling house of MacLochlainn, his succession to the diocese of Derry which lay within their former kingdom and in close proximity to their former stronghold at Ailech might be interpreted as regaining a degree of secular political power by ecclesiastical means. It may also be significant that the Mac Lochlainns survived their political fall 'only as a minor clan in Inishowen' in the vicinity of Derry (ibid.).

60   'Ein Bruchstück der Akten des Konzils von Vienne', in P. Ehrle (ed.), *Archiv für Literatur und Kirchengeschichte*, iv (1888). It would be very interesting to know if Mac Lochlainn was present at the council.

61   TCD, Ms.557/2, pp. 220–1 (Reeves transcripts of the Register of Archbishop Fleming of Armagh, 1404–16), dated at Armagh on 20 November *anno regni regis Edwardi primo*, and in which *Dominum Edwardum Dei gratia Regem Hiberniae* is named as the guarantor of Donal O'Neill's undertakings. In the calendar of Archbishop Fleming's register by H.J. Lawlor in *Proc. RIA*, 30C (1912–13), pp. 142–3, King Edward is identified as Edward II of England and this charter is accordingly assigned to November 1307. However no medieval English king was ever styled as 'king of Ireland' and this is clearly a mistake for Edward Bruce, king of Ireland. The date of the charter must therefore be 20 November 1315, if Professor Duncan's argument that Bruce was inaugurated in 1315 rather than 1316 is accepted ('The Scots' invasion of Ireland, 1315', pp. 109–10).

62   See C. Brady, 'The decline of the Irish kingdom', in M. Greengrass (ed.), *Conquest and Coalescence: the Shaping of the State in Early Modern Europe* (London 1991). Surprisingly the dis-cussion of 'the idea of Ireland's "English Constitution"' (ibid., pp. 96–101) makes no mention of the need to supersede the papal bull *Laudabiliter* of *c*.1155 as a justification for English rule in Ireland once the breach had been made between England and the papacy in the 1530s.

MILITARY COUP D'ÉTAT AND ARMY
MUTINIES IN ENGLAND, 1648—1649

Brian Manning

On 30 November 1648 the New Model Army began its march on London. On 2 December it occupied Westminster and Whitehall and on 6 December a detachment commanded by Colonel Pride excluded from their seats in Parliament those MPs who had voted for negotiations with the King. There followed the trial and execution of the King, the abolition of the monarchy and the House of Lords, and the establishment of the English Republic.

In his eponymous book on Pride's purge David Underdown asks the question whether what happened in December 1648 to January 1649 was a *coup d'état* or a revolution.[1] 'Undoubtedly ... a revolution', says John T. Evans;[2] and Derek Hirst states that '1648–9 brought the only revolution in English history'.[3] Conrad Russell, in an essay published in 1973, argues that there were two revolutions: one in 1642–6 was 'the revolution of Parliament' and would be better characterised as a 'rebellion'; the other in 1647–9 was 'the revolution of the army' and was 'a revolution in the full sense of the term'.[4] Thus Russell dates the real revolution from the beginning of intervention by the army in politics. Mark Kishlansky echoes this: 'The English Revolution began in the summer of 1647.'[5] But Ian Gentles, like Hirst, prefers to date the revolution from 1648 when the army actually seized power and imposed its political programme at Westminster: 'It was not till after the Second Civil War ... that the whole New Model Army became authentically revolutionary. Not before the autumn of 1648 was there a united conviction among all ranks that the Army could only save itself by a political revolution at Westminster.'[6] The actions of the army in December 1648-January 1649 brought about 'the most drastic changes to have occurred in the English state since its appearance', writes Ronald Hutton: 'They thus amply deserve the name of the "English Revolution", in the sense that the landmarks of national politics had been almost completely altered.' And he adds: 'It is a reflection of the discrepancy in evidence, but also, perhaps, of distaste, that historians have tended to focus upon the civilian or parliamentary radicals of the age, rather than these soldiers, the men who made the English Revolution.'⁷

These comments highlight the fact that this revolution was brought about by a military *coup d'état*. Students of revolution generally agree that a *coup* has the following characteristics: it is conspiratorial and prepared in secret; its outcome, whether a success or a failure, is decided swiftly; and it does not involve popular or mass participation but is carried out by a small élite group. The events with which I am dealing do not conform to the first criterion – there was no conspiracy and no secrecy, the army debated and prepared in full public view and gave notice of its intentions – but its actions were decisive and its success swift, and executed by the high command and officer corps without popular or mass participation. Further, students of revolution distinguish between *coups* which are palace revolutions and merely change the persons in power, and *coups* which are revolutionary because they lead to 'significant changes in social or political institutions'. The events of December 1648-January 1649 in England did lead to significant changes in political institutions and so fall into the category of a revolutionary *coup d'état*.⁸

Perez Zagorin in his book *Rebels and Rulers, 1500–1660* observes that in early modern European history *coups* were attempted usually by dissident members of 'noble and aristocratic élites', whereas in the twentieth century *coups* are usually the work of the military.⁹ Thus the events of December 1648-January 1649 in England were unusual and precocious in being a military rather than an aristocratic *coup*. S.E. Finer in his study of soldiers in politics says that military intervention 'as we know it today' dates from the French Revolution:

The experiences of Rome, of the medieval Italian city-states …; the activities of such corps as the Mamelukes, the Janissaries, the Streltsi; all these seem to attest the antiquity as well as the perennialism of military intervention in politics. Nevertheless these differ in essential aspects from the military interventions of today … Only the political role of the English New Model Army may be deemed a genuine precursor of what the last century has brought forth. Its social composition; its political organisation; its ideological basis; its corporate political view, whereby the Army Council treated as an independent power with the civilian authorities – all these prefigure the 'revolutionary' armies of the twentieth century. Like the standing armies of the modern era, too, it sought to validate the claim to rule by claiming to be representative of the people, and if not the people, of the Godly: and like them it was drawn irresistibly into rejecting one after another the legislatures that purported to represent the people …¹⁰

While Finer's generalisations require qualifications and modifications, they draw attention to the fact that here we are dealing, not with aristocratic plots and conspiracies so common in the sixteenth and seventeenth centuries, but with a distinctly modern phenomenon – the intervention of the military in politics, a revolutionary army, a military *coup d'état* imposing revolutionary change. Such a *coup* is most likely to occur and to succeed when established institutions are weakened and the location of supreme power is uncertain or disputed. In this case the monarchy was in eclipse, Parliament was unpopular and distrusted, and the location of supreme power was disputed." The *coup* was made by an élite – the high command and officers of the army – but they were not representative of the traditional governing class. Hutton notes:

> For their time, they were an extraordinary group of people. One very unusual thing about them was that so many of their leaders were of relatively humble social origins. Of the total officer corps in 1648, half came from backgrounds so obscure that no information can be recovered about them. Of the remainder, nearly half came from the middle and lower ranks of society, especially from the towns. Only nine per cent of the total had received any higher education, and a sixth of them had definitely been promoted from the ranks. The highest levels of the corps contained more gentry, but also former artisans. This structure is important to remember, because in late 1648 it was the junior officers and the men who were making the political running."

Thus we are dealing not just with the intervention of an army in politics but also with the intervention of people from outside the class that normally dominated politics: the threat of social as well as political revolution was present.

Recent research into military mutinies and army revolts in the early modern period tends to show that soldiers were moved by grievances over pay and conditions of service, and not by politics, ideology or class. Geoffrey Parker says of the mutinies in the Spanish army in Flanders in 1572–1607:

> Like most civilian revolts of the early modern period, the military mutinies of the Spanish army reveal no evidence of any revolutionary purpose or politically conscious agitators. There is not a hint that any mutineers dreamed of overturning the established order, none that they even wished to influence the government towards making peace ... The mutineers were indeed more like strikers: they wanted to receive the wages they had already earned and a formal promise of better conditions of service in the future. (...) Military upheavals sprang initially from economic pressures (delay in the payment of wages at a time of high prices) and not from political or social aspirations ..."

John Morrill shows that the mutinies in English provincial armies in May-September 1646 and April-August 1647 were over pay: 'It is clear that the demands of the soldiers were almost invariably about arrears' in their pay, and 'there is no evidence that any provincial army questioned the political structures'.'⁴ Ian Gentles concludes, in an article published in 1976, that 'arrears and associated material grievances, much more than ideology, were the engine behind the revolt' of the New Model Army against Parliament in 1647.'⁵ Such work has bred an assumption so powerful that when Bernard Capp came to study the revolt of the seamen of Parliament's navy in 1648 (a counter-revolutionary, not a revolutionary, movement) he thought 'it would be natural to find grievances over pay and conditions' behind it, but he discovered instead that it was 'strongly political in nature'.[16] Gentles himself has revised his views on the revolt of the New Model Army in 1647 and now thinks that it was 'already thoroughly politicised' by that time.[17]

The question I want to explore in relation to the military *coup d'état* of 1648–9 and the army mutinies of 1649 is the extent to which the soldiers were motivated by grievances over pay and conditions, and the extent to which they were influenced by political ideology, specifically the ideas of the Levellers. Immediately before and during the *coup* the only evidence of the views of the private soldiers is the petitions of the regiments and garrisons. Such documents present problems because we do not know who drafted them. It seems likely that they were the work of the officers (they were generally signed only by them), although on one occasion there is evidence of the involvement of representatives of the rank-and-file.[18] The petitions may have been read to the men at a rendezvous and approved by acclamation, but we cannot know what they understood by what they heard, nor whether they responded to some points more than to others or approved some points rather than others. In other words, in a petition which demanded their pay *and* called for the trial of the King, were they more concerned with one or the other?

On 11 September 1648 a petition was presented to the House of Commons with, it was claimed, 40,000 signatures. Organised by the Levellers and commanding 'the support of virtually the whole of the radical interest in the nation, far beyond the confines of the Leveller organisation', it said that they had long expected:

That you would have made good the supreme [authority] of the people, in this Honourable House, from all pretences of Negative Voices, either in King or Lords.

That you would have made laws for election of representatives yearly and of course without writ or summons.

That you would have set expresse times for their meeting Continuance and Dissolution: ... and to have fixed an expressed time for the ending of this present Parl.

That you would have done Justice upon the Capitall Authors and Promoters of the former or late Wars ...

That you would have laid to heart all the abundance of innocent bloud that hath bin spilt, and the infinite spoil and havock that hath been made of peaceable harmlesse people, by express Commissions from the King; and seriously to have considered whether the justice of God be likely to be satisfyed, or his yet continuing wrath appeased, by an Act of Oblivion.

It also called for the reform of tithes and of the legal system. The petitioners returned to the House on 13 September and 'became so bold as to clamour at the very Door against such Members as they conceived cross to their Designs; and said they resolved to have their large petition taken into Consideration before a Treaty [with the King]; that they knew no Use of a King or Lords any longer; and that such Distinctions were the Devices of Men, God having made all alike.' Brailsford comments that this petition and demonstration 'cut a channel for the main current of revolutionary opinion in the country, and swept the Army into action.'[19]

Petitions from the regiments and garrisons demanded that the causers of the civil wars be brought to justice, and specifically that the King be put on trial. The influence of the Levellers is evident in expressions of support for their manifesto of 11 September in more than half of the army petitions, some of which underlined the main points of that manifesto, and even went further. The regiment of Colonel Overton visualised an elective monarchy: 'That all future Kings be hereafter elected by the Peoples Representatives upon conditional trust, or without claiming any negative voyce'. Colonel Hewson's regiment suggested a republic, urging 'that the Government of Venice, Holland, Switzerland, and other parts may be examined, that we may not idolize any one Creature, nor never be any more at this charge'.[20] Petitions were invariably in the name of the officers and men but one petition was from the private soldiers alone. This was from the regiments of Colonel Scroope and Colonel Saunders and it advanced a characteristically Leveller doctrine:

concerning the Kings coming in to be King over us, whether by any rule from God or the election of the people, or by the Sword: ... we find no other but that his inheritance came through William the Conquerer by the Sword. Neither do we find in the Law of God any such Command, that this man should be King in England: And experience tells us that he had not his election by the major part of the people of this Kingdom: so that it doth plainly appear, that the Foundation of his being King, was by the Sword: if it be so ... that which gave a being to a King, retains power to dis-inable the same ...'[21]

Victory of the army over the King in the civil war extinguished his claim to the crown which he owed to the Norman Conquest.

The army was also bitterly aggrieved over pay, which had not been increased despite the rise in food prices and which was seriously in arrears. In the autumn of 1648 some regiments had received no wages for five or six months.[22] The officers and soldiers were deeply angered that as a result they had to live at free quarter, which made them a sore burden on the people and greatly hated. The regiments of General Ireton and Colonel Fleetwood protested that 'notwithstanding the unreasonable taxes dayly extorted' from the people 'in the name of pay for the Army', the money did not come to the soldiers, who were thus unable to pay for their lodgings, which was an 'intolerable burthen' on the people and brought odium on the army. The regiments of Colonel Pride and Colonel Deane said that they were

> ear witnesses of the peoples dayly and sad Complaints of the heavy Taxations that lye upon them, more especially, because they see the Kingdoms Treasure perverted to wrong use, and the Souldier unsatisfied, for whom it is pretended to be raised; whereby a heavier burthen then these Taxes lye upon them, by reason of Free-quarter, which proves so great a discouragement to the Conscientious Souldiers, who have cheerfully undergone all other hardships.

Pride's men confessed that 'not an ingenious Souldier of the meanest ranke, but doth blush' to be forced to live at free quarter upon the poor people. The army suspected that it was the deliberate policy of the dominant faction in Parliament to withhold their pay so as to make them unpopular with the people for living at free quarter.[23]

H.N. Brailsford suggests that the soldiers may have been as ready to act against the Parliament, which obliged them to take free quarter, as against the King, who caused the civil wars. When the army purged the Parliament an MP claimed to have heard soldiers say: 'These are the men who have cozened the state of our money and kept back our pay.'[24] Blair Worden thinks that the influence of the Leveller programme of reforms on the soldiers was limited, and that they were primarily concerned to bring the King to justice and to obtain their wages. 'The troops were evidently given to believe that a march on London would provide access to large sums of money stored in the city treasuries, and that the MPs who were to be imprisoned had been largely responsible for withholding their pay.'[25] Certainly Lord General Fairfax wrote to the Mayor and Common Council of London shortly after the occupation of the capital demanding that the city pay its arrears of taxes and immediately supply the army with £40,000.[26]

We do not know what was in the minds of the rank-and-file of the army, perhaps most of them were more concerned over pay than over political issues, but this was not the case with the officers and some of the NCOs and privates, who did not regard pay as sufficient justification for military intervention in politics. Derek Massarella points out that many members of the army,

> especially the officer corps, were volunteers who felt that they *were* fighting for a cause. Indeed, one of the most obvious things about the Civil War is that besides being a struggle for power it was also about issues of principle and that those who took up arms, including many members of the New Model, felt ... that they were involved in a fight to see those issues settled. This is central to understanding why the army became a political force in 1647 and why it remained one.[27]

Mark Kishlansky observes that there was always the potential for the soldiers to revolt over pay, but when the army first intervened in politics, in 1647, although Parliament voted numerous concessions to satisfy the soldiers' material grievances, that did not prevent military intervention because the army denied that it sought merely its own material ends but far more 'the future security of the common right and freedom of the nation'.[28] A grievance over pay would cause protest, demonstration, strike, mutiny, riot, or merely desertion, and remain non-political. Political action needed political motivation and ideological justification, and the army believed that the *coup d'état* of December 1648 was undertaken from the highest motives in the best interests of the nation. Then, as in 1647, the army had a positive political programme for the settlement of the issues raised by the civil war, a programme which owed much to the influence of the Levellers.[29]

The army saw a necessity to work with the minority of civilian politicians in Parliament who were sympathetic with its aims, and with the groups of religious and political radicals in London and the provinces who were the core of the cause for which it had fought, and specifically with the Levellers.[30] The officers discussed with the Levellers the shape of a constitutional settlement.[31] 'The Levellers were an important factor in the calculations of the army leadership and their ideas were treated seriously at the time of the revolution in late 1648 and early 1649'.[32] 'By the closing weeks of 1648', writes Barbara Taft, 'though the King's removal was all but certain, only the Levellers had developed a comprehensive plan for alternative government.' The Levellers held the intellectual initiative from the Petition of 11 September down to 11 December 1648, when they passed to the Council of Officers the proposed new constitution for the nation, entitled the 'Agreement of the People'.[33] John Lilburn, the Leveller leader, objected to the officers amending it and submitting it to Parliament, for he expected or hoped that it would have been sent directly to the people for their approval. In doing so he split his own party and deprived it of a role at this crucial moment of the revolution. Lilburn and Richard Overton walked out of the discussion in the Council of Officers on 14 December, but the other two Leveller spokesmen, John Wildman and William Walwyn, continued in the discussions with the officers until the 18th. On 15 December Lilburn published his own version of the 'Agreement of the People', and when he protested to Fairfax about the procedures and proceedings of the Council of Officers, he was joined by Richard Overton, Thomas Prince and John Harris, but not by John Wildman, William Walwyn, Samuel Chidley and other prominent leaders of the movement.[34] Nevertheless, for five weeks the Council of Officers studied and debated the 'Agreement of the People' and adopted most of the Levellers' ideas and programme, and had the officers' version of the 'Agreement' been implemented the Levellers, including Lilburn and Overton, would not have opposed it.[35]

One hundred and twenty-four officers participated in these discussions (10 general staff officers, 20 colonels, 11 lieutenant-colonels, 13 majors, 56 captains, 14 subalterns) and they were advised by thirty clergy. Barbara Taft finds that the leading part was played by a core of twenty-one officers ranging from generals to captains. Of these one came from the greater gentry, four (possibly five) from the lesser gentry, two had professional backgrounds, and six 'were from merchant families or were themselves small tradesmen': 'ranging from Richard Deane, whose great-uncle had been lord-mayor of London, through Thomas Harrison, son of a prosperous butcher, and Robert Tichborne, a linen-draper, to John Okey, a ship chandler'; 'Colonel John Hewson, usually described as "of mean parentage and brought up to the trade of a shoemaker", may have been a cadet of a landed family in Kent', and he was in fact a substantial shoemaker; Captain George Joyce, famous for his arrest of the King at Holmeby in 1647, had been a tailor; and the origins of the remaining seven are obscure 'but their careers suggest that they came from families of "the middling sort".' This gives a glimpse of the sort of men who made the revolution: generally they did not come from the old ruling class or élites but predominantly from 'the middle sort of people', from which sprang the most dynamic and radical force in the parliamentarian party during the civil war.[36] Taft and Gentles conclude that the proposals for a constitutional settlement which emerged from the Council of Officers were a victory for the junior officers over their superiors – General Ireton, the most voluble and effective spokesman for the senior officers, was defeated in five out of the seven recorded divisions.[37] 'Thus far the story resembles that of many later revolutions', comments Hutton: 'A cadre of radicals had seized power by armed force, equipped with a blueprint for a new political system'. But the officers were prepared to rule only through civilians.[38]

On 20 January 1649 the officers presented their version of the 'Agreement of the People' to the House of Commons for its consideration and approval. They rejected the Levellers' demand for it to be submitted to the people directly and they thought it important to defer to the authority of Parliament, even to the purged rump of Parliament which they had left in control at Westminster. At the outset of the *coup* they had intended to dissolve the Parliament by force, but they were persuaded to purge it instead in order to continue some cover, however ragged, of parliamentary authority for their action, and to preserve a shred of legality in that by a law of 1641 Parliament could not be dissolved without its own consent. Consistently with this it followed that they presented their constitutional proposals to this 'Rump Parliament' and left it to decide what to do with them. They had fought and risked their lives for Parliament in the civil war and even their *coup* against Parliament became in their eyes a *coup* against the 'corrupt' part of Parliament, not against the 'pure' part, not against the institution of Parliament. In their address to Parliament on 20 January 1649 they presented the fundamental philosophy which guided the political intervention of the military. They declared that in the Army's Remonstrance of 20 November 1648 they had set out some general principles for the settlement of the nation, including the Levellers' Petition of 11 September, but these being ignored or rejected by Parliament, they had been forced to use 'an extraordinary way of Remedy' and purge Parliament of 'corrupt members':

Now as nothing did in our own hearts more justifie our late undertakings ... then the necessity thereof in order to a sound Settlement in the Kingdom, and the integrity of our intentions to make use of it only to that end: ... that neither that extraordinary course we have taken, nor any other proceedings of ours, have been intended for the setting up of any particular Party or Interest, by or with which to uphold ourselves in Power and Dominion over the Nation, but that it was and is the desire of our hearts, in all we have done ... to make way for the settlement of a Peace and Government of the Kingdom upon Grounds of common Freedom and Safety: And therefore ... We have thought it our duty [to embody the general principles of the Remonstrance of 20 November in] an intire frame of particulars, ascertained with such circumstances as may make it effectively practicable.

Now to prevent misunderstandings of our intentions therein, We have but this to say, That we are far from such a Spirit, as positively to impose our private apprehensions upon the judgments of any in the Kingdom ... much lesse upon your selves ...[39]

As Massarella says: 'the army did not aim at seizing power to install itself and to impose a settlement on the nation'. It sought to legitimise its *coup d'état* by working through civilians and preserving a semblance of parliamentary authority and legality.[40] 'During different moments of crisis – July and October 1647, December 1648 – and against strong sentiments that the people's safety warranted a military take-over, the Army opted for action that adhered to the principle of parliamentary authority'.[41]

The House of Commons said that it would consider the officers' proposals when 'the necessity of Affairs will permit', and it never did permit. If their political objectives were so important to the officers, why did they acquiesce when their carefully worked-out proposals for a constitutional and religious settlement were ignored by a Parliament which consisted of men they had put in power and kept in power with their swords? The answer of Taft and Gentles is that the proposals were essentially the work of the junior officers, that the senior officers had serious reservations, and that there was no great enthusiasm or even interest on the part of the rank-and-file. 'Without their senior commanders' the junior officers 'had no political power in existing institutions. Nor had they any reason to believe that their political aspirations were shared by the soldiers in the regiments.'[42] But this explanation is inadequate because it leaves out of account the unresolved tension at the heart of the ideology of the army, which is central to understanding the course which the revolution now took; and for the same reason Kishlansky's argument that the ideology of the army rested on the principle of parliamentary authority requires qualification.

Were the officers in fact wholly committed to establishing a 'democratic' constitution on the lines of the 'Agreement of the People', in which power would derive from the people and be exercised by their elected representatives in Parliament? There were inconsistent influences at work in the army. On the eve of the *coup* Cromwell's secretary had written from the forces besieging the last royalist stronghold to a friend at the headquarters of the army at St Albans:

I am very glad and so [are] the rest of our friends, to hear of a beginning of action with you. I verily think God will break that great idol the Parliament, and that old job-trot of government of King, Lords, and Commons. It is no matter how nor by whom, sure I am it cannot be worse if honest men have the managing of it – and no matter whether they be great or no ... The Lord is about a great work ...[43]

The assertion by Gentles that both the officers and the Levellers would have established something like a dictatorship of the 'godly' is misleading.[44] The Levellers' exclusion from the franchise of everyone who would not sign the 'Agreement of the People' is not the same as restricting the vote to the 'godly'. How broadly or narrowly the Levellers defined 'the people', they did not make 'godliness' a qualification for exercising the franchise or holding office. Their concept of 'the people' embraced both saints and sinners. Gentles blurs the very real distinction between the principles of the Levellers and the concept of 'the rule of the saints', and fails to take account of the reservations about the Levellers' proposals in the minds of the 'godly', including 'godly' officers. Colonel Harrison's support for the officers' version of the 'Agreement of the People' in January 1649 was ambiguous and qualified. On the one hand he thought it desirable that the army 'hold forth [to the nation] those things [which tended to] a settling' of things of concern to the people, so that the army should not appear to be seeking its own ends and to get power into its own hands 'that we may reign over them'.

I think the hand of God doth call for us to hold forth [something] to this nation, and to all the world, to vindicate the profession that we have all along made to God, [and] that we should let them know that [what] we seek [is] not for ourselves, but for [all] men.
[On the other hand] ... the word of God doth take notice that the powers of this world shall be

given into the hands of the Lord and his Saints ... Now by this [Agreement of the People] we seem to put power into the hands of the men of the world when God doth wrest it out of their hands ...

He accepted that the 'Agreement' would serve a function for the moment, because the time was not yet come, although it was at hand, when God would come 'forth in glory in the world' and put power into the hands of the saints. This 'Agreement', declaimed Harrison, was the work of men but 'our agreement shall be from God'.[45] The issue was put clearly in a petition to the officers from 'godly people' in Norfolk: 'How can the Kingdom be the Saints, when the ungodly are electors, and elected to Govern?'[46] Such thoughts found a sympathetic audience in the army, not only amongst some of the officers but also amongst some of the rank-and-file. A manifesto of the army in 1650 proclaimed that its aims were 'the destruction of Antichrist, the advancement of the kingdom of Jesus Christ, the deliverance and reformation of his Church, in the establishment of his ordinances amongst them in purity according to his word, and the just liberties of Englishmen'.[47] There was the rub: how to reconcile 'the advancement of the kingdom of Jesus Christ' with the establishment of 'the just liberties of Englishmen' when all recognised that the 'godly' were a very small minority. S.R. Gardiner long ago pointed out that at the heart of the revolution lay two incompatible ideas – reform by means of the rule of a 'godly' élite over the reprobate masses and reform by means of the rule of 'the people' through their elected representatives in Parliament.[48] This was to prove the tragic dilemma of this and all subsequent revolutions: the conflict between reform from above and reform from below, between authority and liberty.

Many of the officers were members of the congregations of the 'godly' – the radical religious sects. The latter supported the *coup* and the new republican regime in 1649, looking to the army for security against the intolerance of either Anglicans or Presbyterians, and expecting 'the rule of the saints' and the 'godly reformation' of society. The officers were persuaded that the revolution could be secured only by alliance with the 'godly', and most of the sectaries turned away from the Levellers and attached themselves to the officers.[49] The officers did not press for Leveller 'democracy', which might have resulted in the re-establishment of Presbyterian or Anglican control, but neither did they impose direct military rule or erect a military dictatorship, and nor in the end did they secure 'godly rule' or 'godly reformation'.[50]

Finer distinguishes four levels of military intervention in politics: (1) 'Influence' and (2) 'Blackmail', which both involve the military in working behind the scenes on the civil authorities; (3) 'Displacement', when the military by violence or the threat of violence put one set of civilians in power in place of another; (4) 'Supplantment', where the military sweep away a civilian regime and establish themselves in its stead.[1] The *coup d'état* of 1648–9 in England proceeded to the third level but not to the fourth.

This *coup* was the work of the officers: the military mutinies of 1649 were revolts of the other ranks. The question remains how far the ordinary soldiers were moved by grievances over pay and conditions and how far by political dissatisfaction. How did the other ranks react to the failure of the officers and the new republican regime to implement the 'Agreement of the People' in either the officers' or the Levellers' version?

The unity of the army in the *coup* did not entirely silence the usual antagonism between officers and men and this was expressed by those who purported to speak as private soldiers. They said that the officers were well paid and lived comfortably but the common soldiers 'go very poor and thin in habit', have a 'Dogs life, hungry meales, Long marches, and hard Lodgings' in churches and empty houses where they sleep on bare boards. 'An Officer undergoes little or no hardship at all' but the common soldier 'undergoes great hardship with hard labour night and day'.

It is not unknowne unto you what perrills and danger we have under you during the whole warre, how we that are the private souldiers are they who fought and conquered the Kingdome and yet our officers they have reaped the honor and profitt of all our enterprises and sufferings, ... they have been recompensed and rewarded and we continue still in our old condition of want and misery, and if we have gotten but a red coat which is a fools Livery we have thought our selves sufficiently rewarded and recompensed ...

Must we not goe and run when they will have us, and whither they will send us? Must we not lye at their doores day and night like doggs to watch and guard them ...

Are we not scorned and abused, and kicked like dogs by them, as if we were the very scumme of the world in their esteeme?

I appeale to all my fellow souldiers I meane especially the foot ... whether we are not used more like beasts than men, like slaves than Christians and whether we that fight for the freedome and Liberty of the Subject are not the men that ar most subjected to thralldome and slavery, having lost that our selves which we seek and laboure to obtain for others ...[2]

In the autumn of 1648 the resentments of the soldiers were channelled against the Parliament, but now that their commanders were identified with the new political regime, it was the officers who became the targets of the soldiers' grievances over pay and conditions and their disappointments that such hopes as they had of reform were not being realised.[3] The militant faction, led by Lilburn and Overton, which dominated the Leveller movement from February to April 1649, had two bases – the army and London, and two targets – the senior officers, who were alleged to have betrayed the cause, and the members of the Council of State and the 'Rump Parliament'. They pursued a campaign with two main demands: one was for the re-establishment of the General Council of the Army, as it had existed in 1647, consisting of the generals with two officers and two other ranks elected by each regiment, to be the supreme authority in the army;[54] the other was for the implementation of the 'Agreement of the People', the dissolution of the present Parliament and the holding of new elections under a new constitution making Parliament responsible to 'the people'.[55]

The Levellers feared that the country was falling under a military dictatorship. In *The Picture of the Council of State*, which was published early in April 1649, Lilburn asked:

Can those Gentlemen siting at Westminster in the House, called the House of Commons, be any other than a Factious company of men trayterously combined together with Crom. Ireton, and Harrison, to subdue the Laws, Liberties, and Freedomes of England ... and to set up an absolute and perfect Tyranny of the Sword, Will and pleasure ... Whether the Free People of England, as well Soldiers as others, ought not to contemne all these mens commands, as invalid and illegal in themselves, and as one man to rise up against them as so many professed traytors, theives, robbers and high way men, and apprehend and bring them to justice in a new Representative, chosen by vertue of a just Agreement among the People, there being no other way in the world to preserve the Nation but that alone ...

The Buckinghamshire Levellers appealed 'to all our dear brethren in England, and Souldiers in the Army, to stand every one in his place, to oppose all Tyranny whatsoever, and by whomsoever, intended against us.' A handbill distributed in the streets of London on 25 April (perhaps written by John Harris) called on the officers and soldiers and all people concerned about 'the Freedom of the Nation' 'to venture their Lives, and all they have' to oppose the seizure of 'absolute Power' by the 'Great Officers and their Confederates in Parliament, and Councel of State'. 'Therefore keep every man his place and Post, and stir not, but immediately Chuse you out a Councel of Agitators [i.e. agents or representatives] once more to judge of these things: without which we shall never see a new Parliament, or ever be quit of these intolerable Burthens, Oppressions, and Cruelties, by which the People, are like to be beggered and destroyed'.[56] The incitement to disobedience and mutiny, and indeed to violent revolution, was clear enough.

The Leveller leaders were arrested on 28 March 1649 and Walwyn was taken along with Lilburn, Overton and Prince, although he had not been active in the movement for three months, he was thus thrust back into a leading role. His influence became evident during April when the imprisoned leaders turned to seeking an accommodation with Cromwell and the 'godly', and the peaceful implementation of the 'Agreement of the People', of which they published a new version on 1 May. His hand is apparent in the repudiation by the leaders of the use of force to achieve their ends:

We are altogether ignorant, and do from our hearts abominate all designes and contrivances of dangerous consequence which we are said (but God knows, untruly) to be labouring withall. Peace and Freedom is our Designe; by War we were never gainers, nor ever wish to be ...

We aim not at power in our selves, our Principles and Desires being in no measure of self-concernment: nor do we relie for obtaining the same upon strength, or a forcible obstruction; but solely upon that inbred and perswasive power that is in all good and just things, to make their own way in the hearts of men, and so to procure their own Establishment.[7]

How did the soldiers respond to these conflicting signals now coming from the Levellers?

Captain Bray's troop in Colonel Reynolds' regiment of horse mutinied in March 1649 and Captain Savage's troop in Colonel Whalley's regiment of horse in April 1649. Bray's troop had not been paid for nearly six months, they were living at free quarter, and they objected to having their arrears docked by half to pay what they owed for their quarters. But the mutiny was precipitated by the dismissal and imprisonment of their captain. This troop had been raised in Kent and was under the heavy influence of radicals: Captain Bray himself, the Cornet (Christopher Chisman) and the Quartermaster (John Naylier) were Levellers or Leveller sympathisers, and in its manifesto the troop declared its support for the Leveller petition of 26 February, *Englands New Chains Discovered*, which it described as a 'lively portraiture of our apprehensions'. But they failed to win support from the rest of the regiment and they were manœuvred into submission.[8]

The next mutiny occurred when Colonel Whalley's regiment was ordered to move from London to new quarters in Essex. This regiment had been paid regularly, had in fact just received a month's pay, but Captain Savage's troop refused to march, demanding a further two weeks' advance of pay in order to provide for their new quarters in Essex. This was a troop with a long history of religious and political radicalism and the leader of the mutiny was a Leveller, trooper Robert Lockyer. Confronted with a vastly superior force of troops obedient to Colonel Whalley, and with Fairfax and Cromwell bringing up reinforcements, the mutineers fled or surrendered. Lockyer was court-martialled and shot, and his funeral occasioned one of the largest and most impressive of Leveller demonstrations in the capital, involving soldiers, citizens and women wearing the Leveller colours of sea-green. Sources both sympathetic and unsympathetic towards these mutineers agreed that this was simply a dispute over pay and had no political inspiration or objective. The army's official account said that the mutineers 'did not so much as pretend common right and Freedome, nor had the least colour of any grievance lay upon them except their not having their pay before it was either received by their officers for them, or due unto them'. The Leveller account quoted Lockyer as 'much lamenting his most sad condition, being condemned for nothing but asking his pay: And indeed that was the thing which did most trouble him, that so small a thing as contending for his pay, should give his enemies occasion to take away his life; which as he often said, had it been for the freedom and liberties of this Nation for which he had engaged these 7 or 8 years, it would have much added to his comfort'.[59] Both mutinies illustrate how soldiers felt that concern for 'the freedom and liberties of this Nation' was a more legitimate ground for disobedience than a grievance over pay.

The decision to send an expedition to reconquer Ireland, and the selection of the regiments for that service by lot, caused a crisis of conscience in the army. In 1647 the army had refused to go to Ireland or to disband until its demands were met to the satisfaction of a General Council of the Army composed of the generals with two officers and two other ranks elected by each regiment. These demands concerned arrears of pay and indemnity for illegal actions committed by soldiers in performing their military service, but also the 'establishment of common and equal right, freedom, and safety' for the whole nation. Thus the question arose in 1649 whether the army was still bound by this and whether the soldiers should go to Ireland unless a General Council representing all ranks approved the terms for that service and freedom was secured in England. The regiments selected by lot for Ireland were offered two months' pay before embarkation, but Fairfax stated repeatedly that no soldier would be forced to go, though if they would not do so they would be discharged from the army with two weeks' pay and forbidden to re-enlist for six months. It is important to stress that virtually no one in the army opposed the Irish expedition on principle or thought it anything but a necessary and honourable service, but many soldiers thought it extremely unfair to give them a choice between accepting the Irish service with two months' pay or being cashiered with two weeks' pay.[60]

One of the regiments of foot selected for service in Ireland was that of Colonel Hewson and on about 25 April it marched out of London to Romford in Essex. There the colonel put to them the terms for the Irish service and three hundred men, between a quarter and a third of the regiment, refused and were cashiered with two weeks' pay.[61] This provoked the distribution of an inflammatory handbill in the streets of London:

All worthy Officers and Souldiers, who are yet mindful, That you Engaged not as a meer Mercenary Army, hyred to serve the Arbytarie ends of a Councel of State; but took up Arms in Judgement and Conscience, in behalf of your own, and the Peoples just Rights and Liberties. You may now see plainly by the proceedings of Colonel Hewson with his Regiment, That the Design of our Grand Officers is, To reduce the Army to a meer Mercinary and Servile temper, that shall obey all their commands, without so much as asking a question for conscience sake ... Colonel Hewson fals to work and Disbands all those Souldiers and Officers that refused to Engage in the Service of Ireland, before the Rights and Liberties of England (which were never more trod under foot) be restored to the People. The end of this, being to be a Leading Case to all other Regiments, both Horse and Foot ... Therefore keep every man his Place and Post, and stir not, but immediately Chuse you out a Councel of Agitators [i.e. agents or representatives] once more to judge of these things ...[62]

News of what had happened in Colonel Hewson's regiment agitated Colonel Scroope's regiment of horse, which had left London on 5 April and now was quartered at Salisbury and Malmesbury in Wiltshire. They 'fell into serious debate ... whether we could lawfully, in safety to our selves, and our Native Rights in England, submit unto that Forraign Service, or no?' Each troop elected 'agitators' who drafted resolutions which were put to a mass meeting of the soldiers on 1 May and approved by the great majority of the other ranks and a handful of junior officers. They printed their resolutions and published them for the consideration of the whole army.[63] Ireton's regiment of horse, which had a radical tradition, was stationed in West Sussex, and when they saw the manifesto of Scroope's men, four troops refused to obey their officers and rode off to Salisbury. A joint declaration of the six troops of Scroope's regiment and four troops of Ireton's regiment protested at the attempt to compel them by 'unequall terms' to undertake the Irish service, 'so as if we should deny, to be presently cashiered the Army, with little or no pay at all in hand, whereby we must either be forced to beg, steal or sterve'.[64] They moved from Salisbury, through Marlborough, to Wantage where they were joined by two troops of Colonel Harrison's regiment of horse (interestingly not assigned to Ireland) bringing their total strength to about 900 men. Quartering at Burford in Oxfordshire they were surprised and overwhelmed by a force of five loyal regiments brought from London by Fairfax and Cromwell. Cornet James Thompson, Corporal Church and Corporal Perkins were shot for mutiny.[65] These dissident troops had focused throughout on a single demand: they would not go to Ireland, and not be divided or disbanded, but would stand on their guard until the General Council of the Army was reinstated, with elected representatives of the ordinary soldiers as well as of the officers, and freedom was secured in England. When that was done they would go to Ireland.[66]

Concurrently with the mutiny at Salisbury, William Thompson, a Leveller who had been a corporal in Colonel Whalley's regiment, and indeed in the same troop as Robert Lockyer, and whose brother was one of the leaders of the Salisbury mutineers, raised the standard of revolt at Banbury in Oxfordshire on 6 May. Lilburn had been associated with him since 1647 and, with John Harris, had stood bail for him in March 1649.[67] Thompson gained the support of two troops of Colonel Reynolds' regiment of horse and of a troop of the Oxford county horse, a total of about 3–400 men. The lieutenant and several troopers of the Oxfordshire horse were Levellers and at the beginning of the year this troop had declared support for the 'Agreement of the People'. The rebels at Banbury subscribed a declaration proclaimed by Thompson and took their stand upon the new version of the 'Agreement' published by the Leveller leaders on 1 May:

We are gathered and associated together upon the bare accompt of Englishmen, with our Swords in our hands, to redeem our selves and the Land of our Nativity, from slavery and oppression ... We will endeavor the absolute setlement of this distracted Nation, upon that forme and Method by way of an Agreement of the People, tendered as a Peace offering by Lieut. Col. John Lilburn, M. Will. Walwyn, M. Thomas Prince, and M. Richard Overton, bearing date May 1. 1649, the which we have annexed to this our Declaration as the Standard of our Engagement, thereby owning every part and particular of the Premisses of the said Agreement, Promising and Resolving, to the utmost hazard of our Lives and Abilities, to pursue the speedy and full Accomplishment thereof ... And particularly, for the Preservation and Deliverance of Lieutenant Colonel John Lilburn, M. William Walwyn, M. Thomas Prince, M. Richard Overton, Captain Bray, and M. William Sawyer, from their barbarous and illegal Imprisonments ... And having once obtained a New Representative, according to the said Agreement ... We shall then freely lay down our Arms ...[68]

At Banbury Thompson's forces were dispersed by Colonel Reynolds with that part of his regiment which had remained obedient to him. Thompson escaped with a dozen followers and took possession of the town of Northampton. There he seized the magazine of arms and ammunition, appropriated the excise money and gave it to the poor, and at the market cross 'read his Declaration, and made a speech to those that came about him, that he would free them from Excise, Free Quarter, Taxes, and Tythes'. For forty hours he 'kept that walled Town in awe' with a dozen men. He withdrew in face of the vastly superior forces coming against him: he was hunted down and clubbed to death.[69]

The authorities made clear what they thought most concerned the soldiers when Parliament on 12 May responded to the mutinies by increasing the pay of the army.[70] But it is not possible to dismiss the mutinies in Scroope's and Ireton's regiments as being merely about pay, as Hutton does.[71] Were they Leveller mutinies? C.H. Firth and Godfrey Davies claim that they were 'instigated by the Levellers': 'The political demands of the mutineers were extravagant. Inspired by the pamphlets of Lilburn and other Levellers, they demanded the immediate realisation of the democratic programme set forth in the Agreement of the People'. And Christopher Hill says that the mutinies in these regiments turned into 'a full-scale Leveller revolt'.[72] But Brailsford correctly points out that the mutineers in Scroope's and Ireton's regiments did not mention the 'Agreement of the People' nor the imprisonment of the Leveller leaders.[73] Since these were the two issues on which the Levellers were campaigning most intensively at this time, their omission tells against Leveller influence; although one report did relate that some of the mutineers 'spake dissatisfaction at the proceedings against Mr. Lockyer, and of the strict dealing with the prisoners in the Tower'.[74] Gentles, nevertheless, does describe these mutineers as Levellers: 'I use the term "Levellers" advisedly. Although there is no positive proof that the London Levellers organised the mutiny of Scroope's, Ireton's and Harrison's men, and while the mutineers made no public reference to the Agreement of People of 1 May 1649, their programme was clearly Leveller-inspired'.[75] A distinction must be made, however, between the mutineers at Salisbury and the rebels at Banbury, for the latter did take their stand overtly and primarily upon the Levellers' 'Agreement of the People' of 1 May 1649. Lilburn denied any involvement with the mutineers at Salisbury and criticised them for not declaring for the 'Agreement'.[76] Unlike Thompson they were not seeking to launch a violent revolution. Their action was intended to draw attention to their grievance against the terms for the Irish service and to enter into negotiations with Fairfax for its redress. They said that they were 'unwilling to shed blood, or to be the original occasion of a new war' but their resolution was 'to stand on our guard and capitulate [i.e. treat] with our swords in our hands.' They did try to negotiate with Fairfax.[77] Lilburn condemned this tactic: 'when they do draw their Swords against their General, Etc. they shall throw away their Scabbards, and rather fight with him, then treat with him, without either resolving to give or take Quarter'.[78] That was precisely Thompson's spirit, although Lilburn remained silent about his action.

However, the demand of Scroope's and Ireton's men for the reinstatement of the General Council of the Army was one which the Levellers had been making since March.[79] This was a military grievance, and the Salisbury mutineers confined themselves to things which affected them as soldiers, but it did have a political dimension, because the General Council was to pronounce on what was needed to secure the peace and freedom of the nation. The complaint about the terms for the Irish service was politicised and led the mutineers to claim the right of the soldiers to participate in politics:

Wherefore we are now resolved no longer to dally with our God, but with all our endeavours to pursue what we have before promised in order to the setling of this poor Nation, and the restitution of our shaking Freedom, and redeeming our selves out of the hands of Tyrants; for which cause (the safety of the Nation involved together with our own) hath forced us to deny obedience to such Tyranicall Officers whose unsufferable proceedings tend manifestly to the obstruction of our Peace, the hinderances of the Relief of Ireland, the re-inslaving of the consuming Nation ... And to the People of this Nation We Declare, and Protest, We seek not the Ruine but Continuance of Parliaments, We seek ... your Freedom from those intollerable burthens lying on your Shoulders, whereof we are very sensible ...[80]

This fell short of endorsement of the Leveller programme or support for the Leveller leaders. It is likely that this was the minimum on which all the mutineers could agree, and that some of them would not declare for the 'Agreement' and the Levellers. In any case they left the questions of peace and freedom to the judgment of a revived General Council of the Army and it would have been inconsistent to have prescribed in advance that it must adopt the full Leveller programme. Henry Denne, one of the leaders of the mutiny, said: 'we were an Heterogenial Body, consisting of parts very diverse one from another, setled upon principles inconsistent one with another.'[81] This remark has often been quoted as if it alluded to the Leveller movement but in fact it referred to the Salisbury mutineers. No doubt the Leveller agitation during the previous weeks, and individual Levellers amongst the soldiers, influenced the mutinies in Scroope's, Ireton's and Harrison's regiments, but it is misleading to call them a Leveller revolt. They were, however, not just about pay and conditions for the Irish service, they had also a political and ideological focus, without being specifically or overtly Leveller.

After a lull Leveller activity intensified again during August and September 1649. But the movement was divided again. One group urged the soldiers to try to get the support of their offices and to petition Fairfax, and said 'there needs no violence we hope at all'. Lilburn, however, rejected this: 'I cannot advise you to make address to him, as the General of the Nations forces, for he is no such thing; but is meerly a great Tyrant, standing by the power of his own will, and a strong sword, born by his vassals, slaves, and creatures'. The 'Agreement of the People' of 1 May was reissued on 21 August and on 23 August 'the oppressed of the County of Surrey' petitioned for it to be implemented speedily. But Lilburn repudiated the traditional Leveller tactic of petitioning Parliament: 'I cannot upon any terms in the world, either with safety, justice or conscience ... give my consent, but hinder as much as I am able, all addresses from you, or any other ... that shall own those usurping Tyrants as a Parliament, especially by Petition'. He identified himself with the ex-soldiers who had participated in the mutinies in Scroope's,

Ireton's and Harrison's regiments. They published on 20 August a defence of the mutineers, entitled *The Levellers (Falsly so called) Vindicated*, and called for a revolutionary convention of representatives elected by each regiment of the army and by each county of England to consider how to accomplish the 'Agreement'. On 29 August *An Outcry of the Youngmen and Apprentices of London* was directed 'to the private Souldiery of the Army' by supporters of the 'Agreement' and of the mutineers:

> We say ... we are necessitated and compelled to doe the utmost we can for our preservations; and for the preservation of the Land of our Nativity; and never (by popular petitions) addresse our selves to the men sitting at Westminster any more, or to take any notice of them, then as of so many Tyrants and Usurpers, and for time to come to hinder (as much and as farr as our poor despised interest will extend to) all others whatsoever from subscribing or presenting any more popular petitions to them ...

They appealed to the private soldiers 'speedily to chuse out from amongst your selves, two of the ablest and constantest faithfull men amongst you in each Troop and Company' to consider with each other and themselves some effectual course to accomplish the 'Agreement'. Lilburn praised these two manifestos as 'the most sensible and choice Peeces, for the present times, that ever I did see with mine eyes', showing 'the true Remedy of all England's Maladies, and the reall-Way to its Peace and Freedome (viz.). The choosing out of Agents amongst those that yet are honest in the Army and Countrey, to promote the calling of a new Parliament, upon the Principles of the Paper called "The Agreement of the People", Dated 1 May 1649'. In September *The Remonstrance of many Thousands of the Free-People of England* protested 'against all those Tyrants and Usurpers now sitting at Westminster, or any of their Acts, Ordinances, Decrees, Orders, or Votes; and Denying any the least Obedience to, or observance of the same till all and every our former Petitions (with the Agreement of us Free People of England) be duely Considered, and fully Granted.' They proclaimed a campaign of civil disobedience: 'In the meantime we utterly deny the Payment of all Taxes, Assessments, Tythes, or being burdened with Free-Quarter'. They concluded: 'We have drawn our Swords, and are Resolved not to put them up again, till we have obtained the things before specified'.[82]

At the time of the mutinies in Scroope's and Ireton's regiments the officers of Colonel Ingoldsby's regiment of foot had written to Fairfax deploring the divisions in the army and the throwing off of discipline by some soldiers, but they added that the way to compose these differences was to bring the present Parliament to a speedy end and to settle the Commonwealth by 'an Agreement made amongst the faithfull People of this Nation' for the beginning and ending of all future Parliaments, and for the removing of grievances. They concluded that if these things were done their men would remain quiet and obedient.[83] This radical regiment was recruited mainly in Buckinghamshire and was stationed in Oxford. On 8–10 September it broke into mutiny, together with some of Colonel Tomlinson's cavalry, involving perhaps 8–900 men. Their grievances were over deductions from their arrears to pay for free quarter and the payment of the rest of their arrears in debentures, which the soldiers were forced to sell at a discount in order to obtain ready cash, sometimes getting as little as 3s or 4s in the pound. The mutiny was led by Sergeant Radman, a Leveller, who had been one of the elected representatives of the other ranks on the General Council of the Army in 1647. A Captain Jones of Dean's Court in St Martin's Lane, Westminster, brought him copies of *An Outcry of the Youngmen and Apprentices of London*, and the revolt was precipitated when the officers searched Radman's quarters and found the copies of this manifesto and other papers 'of the same nature'. The Oxford mutineers directed their own manifesto 'to all the Souldiers of the Army, and to the whole Nation'. They demanded the reinstatement of the General Council of the Army, with representatives elected by the other ranks as well as by the officers, 'as the only means of settling the Nation in freedom and liberty'. They also demanded the ending of the present Parliament and the calling of a new one to be elected 'by the people in generall', the abolition of tithes and the reform of the legal system. They declared that their desires were expressed by the 'Agreement of the People' published by 'our foure friends in the Tower of London', which they described as 'an unparrall'd expedient for the settlement of the Nation'. In the face of resolute action by the officers the revolt crumbled from within before the arrival of the four regiments sent from London under the command of Major-General Lambert to suppress it. Radman escaped but two private soldiers, Biggs and Piggen, were shot for mutiny. The Council of State showed its view of the causes of unrest by ordering, as soon as it heard of the mutiny at Oxford, two weeks' pay to be sent immediately to the army.[84]

The *coup d'état* of December 1648-January 1649 was a movement of the officers; the mutinies of 1649 were movements of the rank-and-file, involving a very few junior officers but with NCOs playing a prominent role. The mutiny of Savage's troop was entirely about pay and had no political aims; the mutiny of Bray's troop was about pay and the dismissal of their captain but did express support for the Levellers; the mutinies in Reynolds' regiment, the Oxford county horse, Ingoldsby's and Tomlinson's regiments proclaimed political and avowedly Leveller aims; the mutinies in Scroope's, Ireton's and

Harrison's regiments were about the conditions for the Irish service but also had political though not overtly Leveller aims. There was a cadre of political radicals in the ranks of the army, generally Levellers, to whom soldiers looked for a lead when discontented with their officers, and they assumed the leadership in all these mutinies. With the notable exception of Ingoldsby's regiment the mutinies took place in cavalry regiments. 'The propaganda of the Levellers infected the cavalry far more widely than the infantry, for the troopers were men of greater independence and better education, and had a livelier interest in political questions than the privates of foot.' 'The men of the cavalry were clearly the sons of yeomen and craftsmen – at the lower end of the scale of "middling sort of people" ... The foot, by contrast, were drawn from the lowest ranks of society'.[85] The cavalry thus belonged to the constituency to which the Levellers appealed and from which they got their support. It is wrong to make too sharp a distinction between material concerns and ideological aims. As in the *coup d'état* so in the mutinies the soldiers had a need to justify and legitimise their actions by reference to moral and political principles. The material and military grievances of the mutineers were shared by soldiers who did not mutiny; in those units which did mutiny it was the influence of individual radicals and Levellers that converted discontent into active revolt, and this was true even in the case of Savage's troop, while in the other cases it politicised the discontent.

In the English Revolution, as in most modern revolutions, success or failure of the revolution depended on the attitude of the armed forces.[86] The political revolution of 1648–9 – the purge of Parliament, the trial and execution of the King, the abolition of the monarchy and the House of Lords, the establishment of the English Republic – was forced through by the army. It did not lead to the democratic revolution of the Levellers – the implementation of the 'Agreement of the People', the abolition of tithes, the reform of the legal system. The Levellers and the radical revolutionaries could have succeeded only if a large part of the army joined the mutinies or refused to suppress them. Gentles calculates that the total number of men involved in the mutinies between the spring and autumn of 1649 was over 2,500 and, although this figure may be too high, even so it is a very small proportion of an army of over 40,000 men.[87] More than ninety per cent of the soldiers did not mutiny, though some came close to it, and the vast majority obeyed orders to crush the mutinies and shoot mutineers. Thus the army determined how far the revolution would go. The officers called for unity in face of the dangers to the new regime from Ireland and Scotland, and dwelt on the fear that if the army became divided the Royalists would revive and begin a new civil war. The officers gave out promotions to the obedient and punishments to the disobedient. The Levellers agitated against the taxes which paid the soldiers' wages, and so the soldiers looked to their officers to obtain their pay and their arrears for them.[88] Their loyalty was shaken when they felt they were not being properly looked after by their officers, as in the conditions for the Irish service or in deductions from their arrears or in paying them with debentures, but their loyalty was not extensively shattered. The mutinies of 1649 were a watershed. Their suppression was a victory for military force and a defeat for the more revolutionary and populist section of the 'middle sort of people'. In the late 1640s the army was a revolutionary force: in the 1650s it became a force for order and stability, inhibiting or repressing radical dissent and suppressing popular resistance.[89]

NOTES

1   David Underdown, *Pride's Purge: Politics in the Puritan Revolution* (Oxford 1971), pp. 2, 3, 5.

2   John T. Evans, *Seventeenth-Century Norwich: Politics, Religion and Government, 1620–1690* (Oxford 1979), pp. 182–3.

3   Derek Hirst, *Authority and Conflict: England 1603–1658* (London 1986), p. 291.

4   Conrad Russell (ed.), *The Origins of the English Civil War* (London 1973), pp. 2–3.

5   Mark A. Kishlansky, 'The Army and the Levellers: the Road to Putney', *Historical Journal*, xxii (1979), p. 795.

6   Ian Gentles, 'Arrears of Pay and Ideology in the Army Revolt of 1647', in Brian Bond & Ian Roy (eds), *War and Society*, vol. i (1976), p. 61.

7   Ronald Hutton, *The British Republic, 1649–1660* (London 1990), pp. 4, 5.

8   Harry Eckstein, 'On the Etiology of Internal Wars', *History and Theory*, iv, no. 2 (1965), pp. 135, 141; Chalmers Johnson, *Revolution and the Social System* (Stanford 1964), pp. 49–56; idem, *Revolutionary Change* (London 1968), pp. 155–6.

9   Perez Zagorin, *Rebels and Rulers, 1500–1660* (2 vols, Cambridge 1982), i, pp. 40–2.

10 S.E. Finer, *The Man on Horseback: the Role of the Military in Politics* (London 1962), pp. 205–6.

11 Ibid., pp. 21, 84–5, 87–8.

12 Hutton, *British Republic*, p. 5.

13 Geoffrey Parker, 'Mutiny and Discontent in the Spanish Army of Flanders, 1572–1607', *Past & Present*, no. 58 (1973), pp. 47, 50.

14 J.S. Morrill, 'Mutiny and Discontent in English Provincial Armies, 1645–1647', *Past & Present*, no. 56 (1972), p. 63.

15 Gentles, 'Arrears of Pay', p. 61.

16 Bernard Capp, *Cromwell's Navy: the Fleet and the English Revolution, 1648–1660* (Oxford 1989), pp. 25–6.

17 Ian Gentles, *The New Model Army in England, Ireland and Scotland, 1645–1553* (Oxford 1992), pp. 150, 480 n. 70.

18 Underdown, *Pride's Purge*, p. 119.

19 Don M. Wolfe (ed.), *Leveller Manifestoes of the Puritan Revolution* (reprinted, London 1967), pp. 279–90; William Haller & Godfrey Davies (eds), *The Leveller Tracts, 1647–1653*

(reprinted, Gloucester, Mass. 1964), pp. 147–55; Murray Tolmie, *The Triumph of the Saints: the Separate Churches of London, 1616–1649* (Cambridge 1977), pp. 176–7; H.N. Brailsford, *The Levellers and the English Revolution* (London 1961), pp. 349–57.

20 *The Kingdomes Weekly Intelligencer* (31 Oct.-7 Nov. 1648), British Library E.470(10); *The Moderate* no. 17 (31 Oct.-7 Nov. 1648); *A Petition from Severall Regiments of the Army, viz. Colonell Fleetwoods, Colonell Whalies, Colonell Barksteads* (London 1648), BL E.470(32); *Severall Petitions Presented to His Excellency the Lord Fairfax by the Lieut. Generals, Col. Harrisons, Coll. Prides, Coll. Deanes, Regiments* (1648), BL E.474(5); *A Remonstrance or Declaration of the Army* (London 1648), BL E.473(3); *Two Petitions Presented to His Excellency the Lord Fairfax. The One by the Officers and Soldiers of the Garrisons of Newcastle, Tinmouth, Hartlepoole, Holy-Isle: Together with several Officers of Barwick then present. The other by the Officers and Soldiers commanded by Colonel Hewson* (London 1648), BL E.473(23); *The humble Petition and Representation of the Officers and Souldiers of the Garrisons of Portsmouth, Southsea Castle, Southton, Hurst Castle, Poole and Castle, Weymouth, The Castles, Forts and Forces in the Isle of Wight, and the Garrison of Malmesbury* (1649), BL 669 f.13(73); *A Proclamation of his Excellency: Tho. L. Fairfax, L. Gen.* (London 1649), BL E.537(36); *The Declaration of the Officers of the Garrison of Hull* (London 1649), BL E.545(17); Gentles, *New Model Army*, p. 268.

21 *A Moderate and Cleer Relation of the private Souldierie of Colonell Scroops and Col. Sanders Regiments* (London 1648), BL E.476(25). See also *The Humble Representation of the Desires of the Officers and Souldiers in the Regiments of Horse, for the County of Northumberland* (1648), BL E.475(13).

22 Gentles, *New Model Army*, pp. 239–41; Hutton, *British Republic*, p. 7.

23 *The Copies of two Petitions from the Officers and Souldiers of Col. Charles Fleetwoods Regiment* (London 1648), BL E.468(32); *A Petition from Severall Regiments of the Army, viz. Colonell Fleetwoods, Colonell Whalies, Colonell Barksteads* (London 1648), BL E.470 (32); *Severall Petitions Presented to His Excellency the Lord Fairfax by the Lieut. Generals, Col. Harrisons, Coll. Prides, Coll. Deanes, Regiments* (1648), BL E.474(5); *The Moderate Intelligencer* (7–14 Dec. 1648), BL E.476(24); *The True Copy of a Petition Promoted in the Army, And already presented to His Excellency the Lord General, By the Officers and Soldiers of the Regiment under the Command of Commissary General Ireton* (London 1648), BL E.468(18); *The Declarations and Humble Representations of the Officers and Souldiers in Colonel Scroops, Colonel Sanders, Colonel Wautons Regiments* (London 1648), BL E.475(24); *The Perfect Weekly Account* (29 Nov.-6 Dec. 1648), BL E.475(20); *Two Petitions Presented to His Excellency the Lord Fairfax. The One by the Officers and Soldiers of the Garrisons of Newcastle, Tinmouth, Hartlepoole, Holy-Isle: Together with Several Officers of Barwick then Present. The Other by the Officers and Soldiers Commanded by Colonel Hewson* (London 1648), BL E.473(23).

24 Brailsford, *The Levellers*, pp. 336, 373.

25 Blair Worden, *The Rump Parliament, 1648–1653* (Cambridge 1974), p. 76.

26 *The Demands and Desires of His Excellency the Lord General Fairfax* (London 1648), BL E.475(36); *The Kingdomes Weekly Intelligencer* (19–26 Dec. 1648), BL E.536(5); S.R. Gardiner, *History of the Great Civil War, 1642–1649* (4 vols, London 1893), iv, pp. 273–4.

27 Derek Massarella, 'The Politics of the Army and the Quest for Settlement' in Ivan Roots (ed.), *'Into Another Mould': Aspects of the Interregnum* (Exeter Studies in History no. 3) (Exeter 1981), pp. 42, 46.

28 Mark A. Kishlansky, 'Ideology and Politics in the Parliamentary Armies, 1645–9' in J.S. Morrill (ed.), *Reactions to the English Civil War, 1642–1649* (London 1982), pp. 178–9.

29 Hutton, *British Republic*, pp. 8–9; J.S. Morrill, 'The Army Revolt of 1647' in A.C. Duke & C.A. Tamse (eds), *Britain and the Netherlands* (The Hague 1977), vol. vi, p. 71; Gentles, *New Model Army*, pp. 214–7.

30 C.H. Firth (ed.), *The Memoirs of Edmund Ludlow* (2 vols, London 1894), i, pp. 203–13.

31 Haller & Davies, *Leveller Tracts*, pp. 415–23.

32 Massarella, 'The Politics of the Army', pp. 46–7; see also Gentles, *New Model Army*, pp. 289, 292.

33 Barbara Taft, 'The Council of Officers' Agreement of the People, 1648–9', *Historical Journal*, xxviii (1985), pp. 170–1.

34 Haller & Davies, *Leveller Tracts*, pp. 157–60, 225, 283–4, 423–7; Wolfe, *Leveller Manifestoes*, pp. 294–303; *A Plea for Common-Right and Freedom* (London 1648), BL E.536(22); Tolmie, *Triumph of the Saints*, pp. 180–1.

35 Haller & Davies, *Leveller Tracts*, pp. 283–4.

36 Taft, 'Council of Officers', pp. 173–5; Christopher Hill, *God's Englishman: Oliver Cromwell and the English Revolution* (London 1970), pp. 101–2.

37 Taft, 'Council of Officers', pp. 175–7, 183, 185; Gentles, *New Model Army*, p. 291.

38 Hutton, *British Republic*, p. 10.

39 Wolfe, *Leveller Manifestoes*, pp. 334–6.

40 Massarella, 'The Politics of the Army', pp. 44, 47, 48.

41 Kishlansky, 'Ideology and Politics', p. 176.

42 Taft, 'Council of Officers', p. 184; Gentles, *New Model Army*, pp. 293–4.

43 'Leyborne-Popham Manuscripts', *Historical Manuscripts Commission* (London 1899), vol. li, pp. 8–9.

44 Gentles, *New Model Army*, pp. 292–3.

45 A.S.P. Woodhouse (ed.), *Puritanism and Liberty: Being the Army Debates (1647–9)* (London 1938), pp. 177–8.

46 *Certain Quaeres Humbly Presented In way of Petition* (London 1649), BL E.544(5).

47 Woodhouse, *Puritanism and Liberty*, pp. 475–6.

48 S.R. Gardiner, *History of the Commonwealth and Protectorate* (4 vols, London 1903), i, p. 29.

49 Tolmie, *Triumph of the Saints*, pp. 188–9; Bernard Capp, *The Fifth Monarchy Men* (London 1972), p. 230; Gentles, *New Model Army*, pp. 416, 435.

50 Worden, *The Rump Parliament*, pp. 78–80; Austin Woolrych, 'The Cromwellian Protectorate: a Military Dictatorship?', *History*, lxxv, no. 244 (1990), pp. 207–31; Derek Hirst 'The Failure of Godly Rule in the English Republic', *Past & Present*, no. 132 (1991), pp. 33–66.

51 Finer, *Man on Horseback*, pp. 86–7.

52 *A Moderate and Cleer Relation of the private Souldierie of Colonell Scroops and Col. Sanders Regiments* (London 1648), BL E.476(25); *Pay provision and good accomadation for the privat Soldiers* (in manuscript), BL E.537(8); *The Souldiers Demand* (Bristol 1649), BL E.555(29); Wolfe, *Leveller Manifestoes*, pp. 372–5.

53 Gentles, *New Model Army*, pp. 315–6.

54 Wolfe, *Leveller Manifestoes*, pp. 360, 363–4, 381, 383; Haller & Davies, *Leveller Tracts.*, p. 189; *The English Souldiers Standard* (1649), BL E.550(1).

55 Haller & Davies, *Leveller Tracts*, pp. 157–70, 189, 206, 209–11, 225.

56 Ibid., p. 206; G.H. Sabine (ed.), *The Works of Gerrard Winstanley* (reprinted, New York 1965), p. 639; (single sheet printed on one side without title), BL E.551(21), and reprinted in *Mercurius Militaris, Or The People's Scout* (17–24 April 1649), BL E.551(13), which was edited by the Leveller John Harris.

57 Haller & Davies, *Leveller Tracts*, pp. 282, 284.

58 John Naylier, *The New made Colonel Or Irelands Jugling Pretended Reliever* (London 1649), BL

E.552(10); John Naylier & John Marshall, *The Foxes Craft Discovered* (1649), BL E.549(7); Christopher Chisman, *The Lamb contending with the Lion* (1649), BL E.563(10).

59   *Perfect Occurrences* (20–7 April 1649), BL E.529(21); *England's Moderate Messenger* (23–30 April 1649), BL E.529 (25); *A Perfect Summary of an exact Dyarie of some Passages of Parliament* (23–30 April 1649), BL E.529(24); *The Impartiall Intelligencer* (25 April-2 May 1649), BL E.529(29); *Mercurius Pragmaticus* (24 April-1 May 1649), BL E.552(15); *The Moderate* no. 42 (24 April-1 May 1649); *The Kingdomes Weekly Intelligencer* (24 April-1 May 1649), BL E.552(21); *The Moderate Intelligencer* (26 April-4 May 1649); *Perfect Occurrences* (27 April-4 May 1649), BL E.529(32); *A Modest Narrative of Intelligence* (28 April-5 May 1649), BL E.553(1); *The Army's Martyr* (London 1649), BL E.552(11); *A True Narrative of the late Mutiny* (London 1649), BL E.552(18);*The Army's Martyr: or A more ful Relation of the barbarous and illegal Proceedings of the Court-Martiall at White-Hall Upon Mr. Robert Lockier* (London 1649), BL E.554(6), and continued under the same title in E.554(11); *The Justice of the Army against Evill-Doers Vindicated* (London 1649), BL E.558(14); C.H. Firth & Godrey Davies, *The Regimental History of Cromwell's Army* (2 vols, Oxford 1940), i, pp. 218–20. The statement in the entry for Robert Lockyer in Richard L. Greaves & Robert Zaller (eds), *Biographical Dictionary of British Radicals in the Seventeenth Century* (3 vols, Brighton 1982–4), that this mutiny arose when Savage's troop 'refused to suppress Leveller mutineers in Essex' is incorrect and appears to have come from a misreading of Firth & Davies.

60   *Perfect Occurrences* (23–30 March 1649), BL E.529(3); *The Kingdomes Faithfull and Impartiall Scout* (23–30 March 1649), BL E.529(4); *The Moderate* no. 37 (20–7 March 1649); *The Impartiall Intelligencer* (28 March-4 April 1649), BL E.549(15); *A Modest Narrative of Intelligence* (14–21 April 1649), BL E.551(9); *A Perfect Diurnall* (16–23 April 1649), BL E.529(18); *The Perfect Weekly Account* (18–25 April 1649), BL E.552 (2); *A Perfect Diurnall* (30 April-7 May 1649), BL E.529(34); *The Moderate Intelligencer* (19–26 April 1649), BL E.552(4); *A Declaration from His Excellencie* (London 1649), BL E.556(6); *The Resolutions of the Private Souldiery of Col. Scroops Regiment of Horse* (Salisbury 1649), BL 669 f.14(28); *The unanimous Declaration of Colonel Scroope's, and Commissary Gen. Ireton's Regiments* (1649), BL E.555(4); *A Full Narrative of All the Proceedings betweene His Excellency The Lord Fairfax and the Mutineers* (London 1649), BL E.555(27); Norah Carlin, 'The Levellers and the Conquest of Ireland in 1649', *The Historical Journal*, xxx (1987), pp. 280–3.

61   *The Perfect Weekly Account* (18–25 April 1649), BL E.552(2); *The Moderate Intelligencer* (19–26 April 1649), BL E.552(4); *The Kingdomes Faithfull and Impartiall Scout* (27 April-4 May 1649), BL E.529(31); Firth & Davies, *Regimental History*, ii, p. 408.

62   (Single sheet printed on one side without title) BL E.551(21) and reprinted in *Mercurius Militaris, Or The People's Scout* (17–24 April 1649), BL E.551(13).

63   *The Kingdomes Faithfull and Impartiall Scout* (30 March-6 April 1649), BL E.529(8); *The Levellers (Falsly so called) Vindicated, or the Case of the twelve Troops* (London 1649), BL E.571(11); *The Resolutions of the Private Souldiery of Col. Scroops Regiment of Horse* (Salisbury 1649), BL 669 f.14(28); Henry Denne, *The Levellers Designe Discovered* (London 1649), BL E.556 (11); *The Moderate* no. 44 (8–15 May 1649); *The Impartiall Intelligencer* (9–16 May 1649), BL E.530(8); *England's Moderate Messenger* (7–14 May 1649), BL E.530(5); *The Perfect Weekly Account* (9–16 May 1649), BL E.530 (7); *A Full Narrative of All the Proceedings betweene His Excellency The Lord Fairfax and the Mutineers* (London 1649), BL E.555(27); Firth & Davies, *Regimental History*, i, pp. 109–14.

64   *The Kingdomes Faithfull and Imperiall Scout* (4–11 May 1649), BL E.530(2); *The Moderate* no. 44 (8–15 May 1649); *A Modest Narrative of Intelligence* (5–12 May 1649), BL E.555 (8); *The Kingdomes Weekly Intelligencer* (8–15 May 1649), BL E.555 (18); *Perfect Occurrences* (4–11 May 1649),

BL E.530(1); *Mercurius Elencticus* (7–14 May 1649), BL E.555(9); *The Moderate Intelligencer* (10–17 May 1649), BL E.555(25); *The unanimous Declaration of Colonel Scroope's, and Commissary Gen. Ireton's Regiments* (London 1649), BL E.555(4); *A Declaration from his Excellencie* (London 1649), BL E.555(6); Firth & Davies, *Regimental History*, i, pp. 118–23. The evidence does not support the assertion of Gentles, *New Model Army*, p. 534 n. 90, that only three troops of Ireton's regiment joined the mutineers at Salisbury.

65   *A Full Narrative of All the Proceedings betweene His Excellency The Lord Fairfax and the Mutineers* (London 1649), BL E.555(27); *A Declaration of The Proceedings of His Excellency The Lord General Fairfax, In the Reducing of the Revolted Troops* (Oxford 1649), BL E.556(1); *The Levellers (Falsly so called) Vindicated...*, BL E.571(11).

66   *The unanimous Declaration of Colonel Scroope's, and Commissary Gen. Ireton's Regiments* (1649), BL E.555(4); *A Full Narrative of All the Proceedings Betweene His Excellency The Lord Fairfax and the Mutineers* (London 1649), BL E.555(27); *The Levellers (Falsly so called) Vindicated...*, BL E.571(11); Henry Denne, *The Levellers Designe Discovered* (London 1649), BL E.556(11).

67   Wolfe, *Leveller Manifestoes*, pp. 248–58; Haller & Davies, *Leveller Tracts*, p. 179; *The Clarke Papers* vol. ii, Camden Society, 2nd series (1894), vol. liv, pp. 199–200.

68   *The Moderate* no. 26 (2–9 Jan. 1649); *Perfect Occurrences* (4–11 May 1649), BL E.530(1); *The Kingdomes Faithfull and Impartiall Scout* (4–11 May 1649), BL E.530(2); *The Kingdomes Weekly Intelligencer* (8–15 May 1649), BL E.555(18); *The Moderate* no. 43 (1–8 May 1649); *A Modest Narrative of Intelligence* (5–12 May 1649), BL E.555(8); *The Impartiall Intelligencer* (2–9 May 1649), BL E.529(36); *Englands Standard Advanced* (1649), BL E.553(2) and E.555(7).

69   *A Perfect Diurnall* (21–8 May 1649), BL E.530(21); *Perfect Occurrences* (18–24 May 1649), BL E.530(18); *The Moderate Intelligencer* (17–24 May 1649), BL E.556(12); *A Modest Narrative of Intelligence* (19–26 May 1649), BL E.556(17); *The Moderate* no. 45 (15–22 May 1649); *Mercurius Pacificus* (17–25 May 1649), BL E.556(16).

70   C.H. Firth, *Cromwell's Army* (reprinted with an introduction by P.H. Hardacre, London 1962), pp. 184–5.

71   Hutton, *British Republic*, pp. 17–8.

72   Firth & Davies, *Regimental History*, i, pp. 109–10; Hill, *God's Englishman*, p. 109.

73   Brailsford, *The Levellers*, p. 512; Hutton, *British Republic*, pp. 17–8.

74   *A Full Narrative of All the Proceedings betweene His Excellency The Lord Fairfax and the Mutineers* (London 1649), BL E.555(27).

75   Gentles, *New Model Army*, p. 536 n. 120.

76   Haller & Davies, *Leveller Tracts*, pp. 448–9.

77   *The Levellers (Falsly so called) Vindicated...*, BL E.571(11).

78   Haller & Davies, *Leveller Tracts*, pp. 448–9.

79   Wolfe, *Leveller Manifestoes*, pp. 360, 363–4, 381, 382, 383; Haller & Davies, *Leveller Tracts*, p. 189.

80   *The Unanimous Declaration of Colonel Scroope's, and Commissary Gen. Ireton's Regiments* (1649), BL E.555(4).

81   Denne, *Levellers Designe Discovered*, BL E.556 (11).

82   *Sea-Green and Blue, See Which Speaks True* (1649), BL E.559(1); John Lilburn, *An Impeachment of High Treason against Oliver Cromwel* (London 1649), BL E.568(20); *An Agreement of the People* (1649), BL E.571(10); *Perfect Occurrences* (17–24 Aug. 1649), BL E.532(24); *The Levellers (Falsly so called) Vindicated...*, BL E.571(11); *An Outcry of the Youngmen and Apprentices of London* (1649), BL E.572(13); John Lilburn, *Strength out of Weaknesse* (London 1649), BL E.578(18); *The Remonstrance of Many Thousands of the Free-People of England* (London 1649), BL E.574(15).

83 *A Full Narrative of All the Proceedings betweene His Excellency The Lord Fairfax and the Mutineers* (London 1649), BL E.555(27).

84 *The Representation of Colonell Inglesby's Regiment in the Garrison of Oxford, in the behalfe of our selves and all the Nation* (1649), Bodleian Library, Wood 515(6); *The Moderate*, no. 61 (4–11 Sept. 1649); *The Kingdomes Weekly Intelligencer* (4–11 Sept. 1649), BL E.573(10); *The Moderate Intelligencer* (6–13 Sept. 1649), BL E.573(19); *The Perfect Weekly Account* (5–13 Sept. 1649), BL E.573(20); *Perfect Occurrences* (7–14 Sept. 1649), BL E.533(1); *The Kingdomes Faithfull and Impartiall Scout* (7–14 Sept. 1649), BL E.533(2); *A Modest Narrative of Intelligence* (8–15 Sept. 1649), BL E.573(22); *A Perfect Diurnall* (10–17 Sept. 1649), BL E.533(4); *Calendar of State Papers Domestic 1649–1650*, p. 308; C.H. Firth, 'The Mutiny of Col. Ingoldsby's Regiment at Oxford in September, 1649', *Proceedings of the Oxford Archaeological and Historical Society* (1884), pp. 235–46; Firth & Davies, *Regimental History*, i, pp. 377–80.

85 Firth & Davies, ibid., i, p. 121; Gentles, *New Model Army*, pp. 38–9.

86 T.R. Gurr, *Why Men Rebel* (Princeton, New Jersey 1970), ch. 8; D.E.H. Russell, *Rebellion, Revolution, and Armed Forces* (New York 1974).

87 Gentles, *New Model Army*, pp. 345–6.

88 Morrill, 'The Army Revolt of 1647', pp. 67–76; Hutton, *British Republic*, pp. 13–14; Gentles, *New Model Army*, p. 201.

89 J.S. Morrill & J.D. Walter, 'Order and Disorder in the English Revolution', in Anthony Fletcher & John Stevenson (eds), *Order and Disorder in Early Modern England* (Cambridge 1985), pp. 146–7.

THE FRENCH MIDI (LANGUEDOC AND PROVENCE) AND ROYAL TROOPS, 1600–1660

René Pillorget

The question of army supplies in the seventeenth century is often dealt with in a few sentences, not only in textbooks but also in more detailed studies. We read that 'the troops lived off the country', that they made life hard for the peasants, and, as far as France was concerned, that the first effective army service corps was organised by Louvois, after 1660. All of this, in its broad outlines, is true. But if we take a closer look, if we immerse ourselves in local monographs and in the archives, we realise that each French province is a separate case. I propose to examine the problem of relations between civilians and troops in two of them, Languedoc and Provence, the first situated to the west and the second to the east of the Rhône. Together they undoubtedly constitute a geographical unit: in each we find the same combination of plains and hills of Mediterranean climate given over to the cultivation of corn, olives and grapes, as well as sheep-farming, and of mountains where the characteristic traits of this climate lessen as altitude increases. They also form a historical unit: at the time of the Roman Empire, these two provinces were but one. A linguistic unit: in the seventeenth century, while French was the official language the vast majority of the population spoke a different language, closer to Latin, *provençal* or *occitan*. Another trait is common to both these provinces: they were both *pays d'Etats*, like a certain number of other peripheral provinces. Different from the *pays d'élections*, which were in the centre of the kingdom, their institutions were representative of the three orders – clergy, nobility, Third Estate – who each year negotiated the amount of the *don gratuit* (free gift) with the king's representative – a term which implied that in theory they were not obliged to grant it – who voted it, and had it levied by their agents. However, each of these two provinces had its own specific features. Languedoc was much more extensive than Provence; its ports were not very good, unlike Provence. Finally, whereas the Protestants were very few and far between in Provence, they made up an important minority in the mountainous regions of Languedoc, notably in the Cévennes and even formed the majority in certain towns. Nîmes is the best example of this, and there were localities where the Protestants managed to do away with the Catholic liturgy: for example mass was not said at Privas for sixty years between 1560 and 1620.

Between the beginning of the century and 1660, there are two distinct periods: before and after 1635, the year France joined the war against Spain. Before 1635, the question of relations between the civilian population and troops was raised at the 'community' level (community from the word *commune*), whether village or town – a town being defined, regardless of size or economic activity, as a space enclosed by walls and having responsibilities, at least in principle, as regards its defence. After 1635, this remained true, but the question was also raised at the level of relations between the representative assemblies of these two provinces, and the royal power.

During the first of these two periods, 1600–35, the history of these provinces was sharply contrasted. While in Provence peace reigned only troubled by political and social revolt in Aix in 1630 and by a few isolated incidents, we tend to forget that Languedoc was the scene not only of the taking up of arms by the nobility in 1615 and 1632, but also of anti-Catholic violence both in the towns, as in Nîmes in 1617, and in the country, as in the Vivarais area in 1619. In 1621 Protestants attacked Catholic churches, just as they did around 1560. And lastly, Languedoc experienced three religious wars.[1] The first from spring 1621 until October 1622; the second from 1625 till 1626; and the third from September 1627 till June 1629. The toughness of these three confrontations cannot be stressed too strongly; they took place not only in the mountainous part of the province, but also on the plains around Montpellier, Nîmes, Sommières and Béziers. The royal armies used the usual process in enemy country. They *font le dégât* (or wrought havoc), that is to say they pillaged and burned the properties of the cityfolk, particularly around Castres and Montpellier. In 1622 Sommières when it was taken, bought itself off pillage in exchange for a high war contribution. The same year, the prince of Condé, at the head of the royal army, took Nègrepelisse on 10 June, and had the whole population massacred. When he took Pamiers in 1628, he proceeded to a series of hangings, treating all those who resisted him as rebels to royal authority. When the pious King Louis XIII took Privas he had a hundred of the principal prisoners hanged without a trial, and sent a hundred more to the galleys. Richelieu's letters to the Queen Mother, as early as 28 and 29 May 1629, gave horrifying details: apparently the king wanted to terrorise the Protestant population by an act of exceptional severity.[2]

The attitude of the Estates of the province – all Catholic and presided over by the archbishop of Toulouse – seems to have been significant. They supported the royal troops. In 1621, on the occasion of the siege of Montauban, they sent them corn, wine, gunpowder and even money. When one of the Huguenot leaders, the count of Châtillon, left his party and found himself besieged in Aigues-Mortes by his co-religionists, the Estates decided to help him. Noblemen spontaneously joined the royal troops. And what is more, a local desire for self-defence became manifest. The diocese of Albi – a purely civil subdivision, having no connection with an ecclesiastical diocese – raised and supported at its own expense – in this same year of 1621 – a regiment of infantry and a company of light cavalry in order to protect itself from the expeditions of the Protestants of Castres, Lombers, Montauban, Réalmont, and Saint-Antonin. The relations between the representative assemblies of the province and the royal army were thus planned under the sign of cooperation. However, the Estates expressed protest against the 'excesses' and 'violence' of the soldiers, against the 'unbridled licence of the men of war', which did not always spare the goods and persons of the Catholics, on the contrary these protests were confined to generalities. As things are it is hardly possible to extract from the mass of events, and to analyse, the clashes between the civilian Catholic population – in theory friendly and pacific – and the royal troops. These clashes seem literally drowned in the mass of military events.

But in Provence, a region at peace, these facts are few and far between. They occasioned the drawing up of documents, following investigations and legal proceedings. Only a few garrisons with small numbers were permanently stationed in this province, spread out over particularly important strategic points inland or on the coast, such as Antibes or Toulon. But certainly not at Marseilles: this would have been contrary to its statute and privileges. Moreover, whereas hostilities between France and Savoy still had not finished, for a few weeks of 'undecided peace' while a continuation of the war seemed possible, troops that had left the province had to be brought back to it. The last harvest had been very poor (*misérable*). In Marseilles, as in other towns, the high price of wheat was not put down to natural causes, but to the passage of troops. When on the evening of 24 February (a Saturday) the news of the approach of a regiment was about, the rumour led to panic and the taking up of arms. Doors were hastily closed and watches organised. This nocturnal movement was confused and spontaneous. It was the first step in collective reflexes brought about by the approach of troops.[3] But there was also a further step: to ring the tocsin which causes or increases panic. 'To ring the tocsin at the approach of the king's troops' was a serious fault — that was however committed relatively often.

One might expect that there would be fewer incidents between the civilian population and troops garrisoned in a town than between the civilian population and troops passing through. In fact, the cohabitation of both in the same agglomeration brought with it three types of difficulty.

First, problems concerning the behaviour of the soldiers. Some were arrogant. Others could also be brutal and thought they could do anything because they were armed. Frequent complaints were made about extortions and theft. This was for instance the case at Saint-Tropez in 1608[4] or at Toulon and elsewhere. It sometimes happened that these men enjoyed virtual impunity, the officers being extremely tolerant towards them when off duty. On 11 April 1631, in Aix, a soldier condemned to death by Parliament for having killed the person who lodged him was carried off from the public square on the day of the execution, the executioner being chased and put to death in the very Palais de Justice. It also happened that soldiers were very persistent towards women — which could give rise to particularly bitter conflicts in the coastline communes where the men, sailors or fishermen, were often absent. This brings us to the second problem, that of allowances made to soldiers: bed and the *feu commun*, or common hearth for heating and cooking. The soldier was settled permanently in a home, which could be very unpleasant for the inhabitant. However, local arrangements could be negotiated. Thus, at Toulon in 1633, a compromise was reached. The *feu commun* was abolished, and replaced by the supply of salt and vinegar to the soldiers. The third problem was that of the respective powers of officers and municipal authorities — called *Consuls* both in Provence and in Languedoc — as regards lodgings and allowances. The Consuls of the towns were determined to exercise this legacy of the past; a certain number of prerogatives and responsibilities of a military order. They set much store by them. In the case of a governor being absent, they exercised the lieutenancy of the government. They gave the sergeant-major and the captain the password. The *garde bourgeoise* alone, under their sole responsibility, was to guard the gates of the town.

Incidents could degenerate if the Consuls called on the masses, or even just gave them a free rein, like at Antibes, during the *tumulte et sédition* (turmoil and mutiny) of 1615. The situation could become serious when the conflict involved a large number of men or when it happened in a place of strategic importance. This is what happened in Toulon in May and June 1633. Officers having got their men to occupy the Notre-Dame gateway, it was attacked by the *garde bourgeoise* and the population, leaving one person dead and 52 injured.[5]

The affair gave rise to the sending to Paris of two delegations, one civic, the other military, each carrying a bundle of reports and records, and ended with the trial of an *intendant*, a *commissaire-royal*. To a great extent it satisfied the people of Toulon. It confirmed the fact that common justice was alone concerned with conflicts between inhabitants and soldiers – even though the governor of the province had had a gibbet put up in the main square at Toulon to intimidate civilians.

These three problems – the behaviour of the soldiers, the allowances made to them, and the respective powers of the civic and military authorities – were raised in both of these frontier provinces alike, and the same gradation in the incidents could be observed, consistent with the acceleration that sociologists well know: rumour, panic, outbreak of hostilities.

The last mentioned of these incidents dates from 1633. Two years later, France was at war with Spain, and during the following years, in both provinces, a firm determination for defence could be observed. In 1637 the Spanish undertook the siege of Leucate, then that of the imposing château of Salses, on the frontier between Catalonia and Languedoc. In 1639 they disembarked in the Lérins Islands, just off the Provençal coast. In Languedoc, provincial militia were levied by the archbishop of Narbonne, the bishops of Montpellier, Nîmes, Viviers, Mende, Albi, and by various other nobles, at their own expense. The same thing happened in Provence. And these improvised troops contributed – with but moderate success if one were to be truthful – to the success of the royal troops. In any case, the goodwill of the representatives of these two provinces seemed undeniable – at least when a nearby target, directly affecting them, was concerned. But to give food or money, to lodge troops destined to fight outside France, in Catalonia or in Piedmont, did not seem to them in any way to be a moral duty towards their prince.

During the years 1635–60, the problem of relations between civilians and soldiers tended to get worse. It was stable at a local level. In Provence as in Languedoc, clashes very similar to those we have already cited as examples can be found. But added to these difficulties which arose between the different corps of troops and the different municipal authorities, conflicts developed at a higher level: that of the military high command, and that of the representative assemblies of these two frontier-provinces.

During the Fronde, the difference in the pattern of events was exactly the opposite of what it had been during the years 1615–29. Whereas during the reign of Louis XIII, Languedoc had been the scene of civil war and Provence of peace, during the minority of Louis XIV, in 1648–53, Languedoc was practically calm, and Provence was the scene of civil war – limited both in time and in space. The Aix Parliament and the governor of the province, the count of Alais, who had raised a regiment for himself in the province, confronted one another, first in the streets of Aix, in January 1649, then in the open country. Victorious in the narrow streets of the old town, the improvised parliamentary troops were crushed in the plains, and after that the Provençal Fronde consisted only in little skirmishes and local incidents. This difference between Provence and Languedoc appeared still more clearly after the Fronde, when France again took the initiative of operations against Spain. To be sure, the north of the kingdom remained the major theatre of operations for Mazarin. But he did not for all that neglect the Mediterranean. He would have liked to batter the Spanish defences there. In Naples, Masaniello's rebellion seemed to offer a possibility for action. In Catalonia, the port of Rosas, in the south of Cape Creus, had been occupied by French troops since 1645, and finally others fought the Spanish in Northern Italy. During the three years that followed the Fronde, 1654–6, Mazarin managed to launch a few modest operations, with limited targets. In Naples, the duc de Guise's expedition ended in failure. But elsewhere the action of the French troops was often successful. In Catalonia the duc de Mercœur managed to enlarge the region occupied around Rosas. In Italy he managed to take Valence, after a two month-siege. For Mazarin, Provence and Languedoc were *l'arrière*, the hinterland where troops could be sent to 'cool down', after action. This made many demands on these two provinces, demands for funds which were added to the *don gratuit*, the sum that the representative assemblies granted the king each year.

Since 1641 the province of Languedoc had tried to see that it should not only be the communes or dioceses which the troops passed through that should support this financial burden, but that it should be shared out over the whole of the province. Besides the sums assigned to the *don gratuit*, others were collected throughout the province, grouped together under the name of *étapes* or 'stages' in the coffers of the Estates. The latter had to pay back to the communes through which the troops had passed the food that they had supplied them with – and to compensate them for any damage done. However, the system worked far from perfectly.[6]

In 1643, the high command considered it necessary to have troops transported by sea from Agde and Narbonne right to Catalonia. The result of the collection made on the score of the *étape* was thus allocated to pay for this transport and nothing was left for the communes that had supplied food and money to the troops during their voyage over land to the two ports of embarkation. The Estates then decided that towns and villages would not,.

without their authorisation, have to supply lodgings or services to passing troops. A decision which, if they respected it, risked laying them open to reprisals, and supplementary extortions on the part of the soldiers.[7]

In the communes they crossed and those where they stopped, the officers did not generally want their men to be supplied with food. For reasons easy to understand, they preferred to receive money. The amount to be given per ration was fixed by a regulation. But they required that the number of rations, payable in cash, should be calculated according to the theoretical numbers of their units – and not according to their real numbers, which were always fewer. This was a source of tedious discussions with the Consuls of the communes, who were forced to negotiate their deals with them.

During the years after 1641, the Estates of Languedoc in this way granted annually, besides the *don gratuit* – the normal tax – of 1,500,000 livres, a sum of 1,650,000 livres for 'winter quarters' for the troops. They sent a written protest, in moving terms, so that the rights of the province should be safeguarded in the future. They specified in their deliberations that the subsidy was granted *sans conséquence* (without consequence) and that they were obliged to make these concessions *avec douleur* (painfully) or even *comme malgré eux* (in spite of themselves) and *à regret* (with regret). They also protested against the excesses committed by the troops.[8]

In neighbouring Provence, as far as the towns and villages were concerned, the same problems of relations with the troops are to be found: requests for money rather than allowances in kind, requests for payment for theoretical numbers rather than for real ones. It was at the top level of provincial representation that a difference in behaviour occurred. The Estates of Provence, representatives of the three orders, no longer met. The king summoned only the Assembly of the Communities which represented the Third Estate and was supposed to be the most docile. It was to this Assembly that the king's requests or demands were presented. This was done by a royal commissioner, an old Frondeur, now repented, baron d'Oppéde, whom Mazarin named First President of the Aix Parliament and who became his agent, unconditionally devoted to the central power. D'Oppéde came up against the opposition of the Assembly of Communities, directed by people called the *Procureurs du Pays*: the archbishop of Aix – cardinal Grimaldi, from a very famous family – and the three Consuls of the town of Aix, the capital of the province, a town known to be unruly.

In Provence as in Languedoc, there was no service of *l'étape* centralised at the provincial level. The troops passing through or in their winter quarters put their requests forward to the authorities of the communes where they stayed. They negotiated arrangements with the officers. The Consuls of the different Communes presented to the *Procureurs du pays* and the Assembly of the Communities their bills, together with proof, representing the totality of the sums supplied and also the damage done by the soldiers. After discussion, the Treasurer paid them. Then, the Assembly 'compensated', that is to say it subtracted the total sum of the bills from the sum voted to the king for the following year by right of the *don gratuit*.

Thus, in 1653, the king asked for a don gratuit of 800,000 livres. The Assembly granted it, but deducted 660,000 and the king had to be content with 140,000 livres of ready money. Everything happened as if the Assembly did not want to take into account the fact that they were at war, and did not want the province to pay more taxes than in time of peace. The stakes of the war in progress, the broad lines of Mazarin's grand European policy were, apparently, completely unknown or indifferent to them. In 1654, Mazarin asked the province for an exceptional supplement of 600,000 livres. The Assembly only granted 340,000, and only on the condition that Provence should be exempt from lodging troops. Mazarin agreed, but did not keep his word: at the beginning of the autumn, troops from Italy broke in on the province, although the *Procureurs du Pays* had not given their *attaches* – that is to say their authorisation. Numerous incidents occurred. At Digne, in particular, one person was killed following a quarrel between civilians and soldiers. After the summer of 1656, marked by a bad harvest and an epidemic of plague, things seemed to be heading for an open conflict between the central government, and its devoted agent the First President d'Oppéde, and all the representatives of the province, indignant at the royal demands.

In 1657, the Assembly of the Communities made a concrete proposal: that the troops in their winter quarters should be distributed over all the communes, towns and villages, in numbers proportional to the size of each of them, and should receive allowances in kind – and not in money. In other words, the soldiers would be dispersed, many of them completely isolated, and at the mercy of the civilian population. As d'Oppéde wrote, 'tout déserterait comme à un siège'.

This proposition, ridiculed in high places, was answered by a royal ruling made public on 30 November by the governor of Provence, the duc de Mercœur. It stipulated that the communes should have to supply the troops with payment in cash, a fixed sum per company and per day, whatever the size of these companies – a sum payable every ten days, for one hundred and fifty days, the theoretical length of the winter quarters, even if the length of their stay in the province was very much less. The two ideas clashed. The *Procureurs du pays* considered the payment granted to the troops by the communes an occasional aid. But in Paris it was seen differently: it was considered an extra tax. They knew very well that the sums asked for amounted to much more than the value of the foodstuffs supplied to the troops and consumed by them. As Mazarin wrote: 'The king does not send soldiers simply to be fed and housed, but also to give them the means to get back into a fit state to serve.' What was serious was that this ruling interfered with privileges, the status of the province, and its autonomy. In practice, Provence found itself treated as a *pays d'élection* – no longer as a *pays d'Etats*. This gave rise to a whole series of serious events in 1658 and 1659,[9] in Aix, where the First President d'Oppéde was personally threatened; in Marseilles, Draguignan and in several other localities. Troubles arising from the question of passage and residence of troops, but which were situated within a very varied context that spread over part of the kingdom: assemblies of *gentilshommes* in the West, urban incidents, peasant movements – agitation which spread from Normandy to the Sologne, an abortive attempt at a rekindling of the Fronde, all of which has never been examined as a whole. In Provence, the troubles led to the coming of Louis XIV in person, together with Mazarin and his Court, with troops, between the signature of the peace of the Pyrenees and his marriage at Saint-Jean-de-Luz. In January 1660, he proceeded to a military occupation of Marseilles, to the remodelling of its municipal institutions, to a breach in its ramparts and the erection of a fortress – not for defending the town against an attack from the sea, but to calm any eventual uprising. Finally, the town was to be subject to the common law in fiscal matters: it would henceforth pay taxes to the king, which had practically never happened before.

The historian is faced with a question: while Provence experienced a veritable conflict with the central power over the problem of the lodging of troops, why did Languedoc, between 1654 and 1660, experience but few incidents which did not go beyond local limits? Why was there no conflict between the institutions representing the province and the central power? Was it in fact the effect of a secular tradition of docility of the Estates of Languedoc towards the authorities? Perhaps. But it was above all the consequence of three factors. Firstly, in 1655, the town of Carcassonne rebelled against the gabelle, the salt tax. The revolt was severely put down, and in 1657 the town was deprived of its privileges, a warning of which others took notice. Secondly, during the Fronde, large-scale brigandry developed in the mountainous area of the province, notably in Gévaudan. The Estates and the parliament of Toulouse needed the army to quell it.[10] Thirdly, these years 1653–9, marked by an ever heavier taxation, were also the years of Protestant uprising in several towns: Lunel, Bédarieux, Nîmes, Montpellier, Réalmont, and Florac. These were not uprisings directed against the king, but developed for entirely local reasons: the acquisition of the *consulat*, which implied, besides authority and prestige, a right to a say in the assessment of taxes. Catholics and Protestants clashed again. This awakening of an old enemy brought the Estates and the Parliament of Toulouse still closer to the royal authority. The internal situation of each province determined the behaviour of its representative assembly. The absence of any serious problems within Provence explained its attitude – which showed just how far it was possible, before 1660, to resist royal demands – and brought about a far-reaching event. The consequences of this conflict of the province with the central authority were the coming of Louis XIV to Marseilles, the breach in the wall of the town and the building of a fortress – commemorated by the striking of a medal, as for a victory – and this whole ensemble shows that a new period in the history of France was just beginning.

NOTES

1   Dom Claude Devic & Dom Joseph Vaissete, *Histoire générale du Languedoc*, vol. xi (Toulouse 1889), pp. 914–1041; vol. xii (1889), pp. 63–7, 127–30.

2   Pierre Grillon (ed.), *Les Papiers de Richelieu. Section politique intérieure* (Paris 1980), vol. iv, pp. 340–7.

3   Bibliothèque Méjanes (Aix-en-Provence), Ms.736, p. 405.

4   Archives Municipales, Saint-Tropez, BB6, fol. 285ss, 289ss, 297ss, 315ss, 432ss.

5   Ministère des Affaires Etrangères, Mémoires et documents, France, 1702, fol. 200–1; Archives Communales, Toulon, BB56, fol. 240 v° 258, 262, 268 pièce 4, and BB322, pièce 3; Archives de la Guerre, A1 18 fol. 335 & A1 19 fol. 45 et 62.

6   Devic & Vaissete, *Histoire générale*, vol. xiii (1877), p. 163.

7   Ibid., p. 165.

8   Ibid., pp. 131–287.

9   René Pillorget, *Les mouvements insurrectionnels de Provence entre 1596 et 1715* (Paris 1975), pp. 751–862.

10 Devic & Vaissete, *Histoire générale*, vol. xiii, pp. 351, 365.

Thomas Bartlett

The Catholic question – the question of the repeal of the Penal laws and the extension to Irish Catholics of full civil, political and religious rights – was central to Irish politics in the reign of George III and for some years after. Quite why the Catholic question should have emerged in the 1760s and then dominated the politics of the next seventy years has long puzzled historians.[2] The setting up of a Catholic Committee in the 1750s appears to have had little to do with it: for the next thirty years this committee was a byword for internal division and rivalry, and its major achievement appears to have been the simple fact that it existed at all. Even in the early nineteenth century its internal conflicts were such that it was seriously suggested that its continuance, so far from forwarding the Catholic cause, actually benefited the anti-Emancipation interest. More plausibly, it has been argued that by the 1760s a decent interval had elapsed since the last Catholic challenge to the Protestant state had been crushed at the Boyne (1690) and Aughrim (1691) and that the subsequent (mostly) peaceable behaviour of Irish Catholics made them now eligible for relief. Ireland, for example, had remained quiet, if tense, during the Jacobite rebellion in Scotland in 1745; and the Jacobite threat appeared to have passed altogether with the death of the Old Pretender in 1766. Again, it has been claimed that the growing evidence of Catholic wealth, and the rumoured Catholic fortunes which the Penal laws excluded from productive investment in the Irish economy, allegedly led some Protestants to look more critically at the laws of exclusion.[3] Some historians have pointed to the 'spirit of the times' and argued that the influence of Enlightenment values and ideas, toleration being prominent among these, played a role in leading Irish Protestants to reassess the Penal code: and the growth of a sentimental, inclusionist nationalism in Ireland during the middle decades of the eighteenth century no doubt helped in this reassessment.[4] Certainly there is evidence that by mid-century the passions of the seventeenth century had somewhat receded in both Ireland and England: Irish Protestant assertiveness against what was regarded as English interference would have been unthinkable without a belief that the threat from Irish Catholics was much reduced; and there is strong evidence that by mid-century anti-Catholicism had declined in England, at élite level at any rate.[5] Moreover, divisions within the Protestant governing élite in Ireland and conflicts between British and Irish politicians – both high-lighted by the Lucas affair of the late 1740s and, above all, by the Money Bill dispute of the 1750s – clearly provided both stimulus and opportunity for Irish Catholics to call for an end to the Penal laws in force against them. Finally, some historians have preferred to examine each of the Catholic relief acts on its own and have attempted to explain each in terms of a unique set of circumstances.[6]

All of these explanations and approaches have merit: taken together they help us understand the political climate that permitted a fresh look at the exclusion of the Catholics, as well as shedding light on the passing of the various relief acts. What is missing, however, is any overarching explanation for the emergence of the Catholic question in the early 1760s, its persistence thereafter, and the ebb and flow of the agitation for it. Nor is there any explanation for the urgency of the Catholic issue, the speed with which it dominated the political agenda and the sense of impending crisis that quickly infused all discussion of the subject. It is the contention of this paper that these elements derived directly from the fact that, from the beginning, military considerations and the question of Catholic relief were intimately connected;[7] that it was above all the pressure of war that forced the Catholic question onto the political agenda, and kept it there.

Admittedly, the Catholic question had ever been associated with war and the threat of war: for the first half of the eighteenth century Irish Catholics were seen as Jacobite in sympathy, and thus inherently disloyal; they maintained what amounted to a standing army abroad – the so-called Irish Brigade in the service of France – which recruited clandestinely among Irish Catholics; and when wars did break out – as for example in 1743 at the start of the War of the Austrian Succession – it was normal for extra security precautions to be taken against them.[8] But from mid-century on all of this changed: war would henceforth mean opportunity rather than danger for Irish Catholics: opportunity to parade their loyalty, to draw up addresses of support, and especially to barter recruits for concessions. Military necessity, especially the manpower requirements of the British army, provided the context for the emergence of the Catholic question in the 1760s and its persistence thereafter.

The eighteenth century ('Enlightenment' notwithstanding) was predominantly an age of war: indeed, some have gone so far as to describe Anglo-French rivalry from the time of Louis XIV to the fall of Bonaparte at Waterloo as constituting in effect a second Hundred Years' War. Moreover, warfare and waging war in the eighteenth century were rather different to what had gone before. In this period warfare became both a labour-intensive and a territorially-extensive pursuit, imposing enormous strains on government machinery and on government finances: P.G.M. Dickinson demonstrated some time ago that the 'financial revolution' of the 1690s was made necessary by the strain of war, while John Brewer has argued recently that the apparatus of the British state in the eighteenth century was a response to the pressures of waging war.⁹ None the less, the size of the army (at least in comparison with Britain's continental rivals) remained low in the years following the Treaty of Utrecht, with perhaps 30,000 soldiers (in Ireland and in England) being found sufficient for all purposes. However, beginning with the Seven Years' War, troop levels, scale of operations and diversity of theatre all increased dramatically. Some 120,000 men were needed to fight the Great War for Empire (as Gibson chose to call the Seven Years' War) in Asia, the Americas and Europe; around 150,000 were mobilised for the American War of Independence; while on the outbreak of the French Wars in 1793 some 175,000 were immediately mobilised and this figure had climbed to 300,000 by 1801 and to 750,000 by 1809, with a further 130,000 in the Royal Navy.¹⁰ Moreover, with war being waged simultaneously on several continents, and with horrendous casualties being incurred in the tropical theatres, the British war machine proved insatiable for fresh recruits. Not surprisingly, from the 1760s on, British politicians and generals turned their attention to Ireland where, it was pointed out; 'there was a weapon of war yet untried by Britain, mighty and strong, that ninety out of a hundred of the King's subjects in Ireland were Roman Catholics.'¹¹

Ireland had not been a recruiting ground for the British army in the early eighteenth century. After the experience of James II's 'Catholick designe', a major part of which had involved 'catholicising' the army in Ireland, all were agreed that it would have been foolhardy to permit the enlistment of Irish Catholics, as it were, 'to guard us from themselves'. Moreover, as an additional safeguard it was also declared illegal to recruit Irish Protestants. (Officerships, ensigncies and cadetships were another matter, and throughout the eighteenth century Irish Protestant officers constituted possibly one third of the British officer corps.) However, so far as Protestant recruitment to the lower ranks was concerned, the thinking was that any depletion of Protestant strength in Ireland would weaken the wider 'Protestant interest' there and thus prove counter-productive; in any case, if Protestants were permitted to enlist, wily Papists might pretend to be Protestants, and thus gain access to the army and arms training. It was deemed safest, therefore, to bar all Irish from the armed forces.

In general, this ruling appears to have been enforced, though there were, of course, attempts made to circumvent it. On one occasion some officers dispatched a parcel of illegal Irish recruits to Scotland, clothed them in plaid and on their return to Ireland attempted to pass them off as Scots. But such practices were dangerous and, if caught, the culprits faced punishment. In 1728 a number of officers were suspended for attempting to enlist Irish recruits; and in 1756 Ensign Baggot had 42 recruits from his native Queen's County disallowed, the Lord Lieutenant, the Duke of Bedford, refusing to 'wink at practices of this sort'.[12] By this date, however, the outbreak of war in three continents was making such scruples increasingly redundant. Already, in 1745, it had been found necessary to lift the ban on Irish Protestant recruits to the lower ranks; but it was the demands of the Seven Years' War that brought about a reappraisal of all exclusion bars. Military necessity quickly led to the recruitment of several battalions of those Scottish (Catholic) Highlanders who had fought for the Pretender barely ten years earlier; in Ireland, the recruitment of Protestants was permitted with the proviso that they would not have to serve overseas; and Presbyterians were permitted to obtain commissions in the Militia.[13]

Furthermore, there is evidence that a blind eye was turned to the recruitment of Irish Catholics, particularly when the regiment concerned was destined for foreign parts.[14] In 1760, for example, Bedford received express permission from the King to recruit Catholics for the marines.[15] More open procedures, however, were not yet possible, though the exigencies of war were transforming principle into pragmatism so far as the exclusion of Catholics from the armed forces was concerned.

This transformation was clearly revealed in 1762, when the Catholic Lord Trimleston offered to help recruit Irish Catholics and, more important, suggested a way around the prohibition on them. His proposals were given serious consideration by the British government. In essence, Trimleston offered to raise six regiments of Irish Catholics (about 3000 men) to serve either the Elector of Hanover (i.e. George III) or the King of Portugal. The first option was ruled out by the King on constitutional grounds; the second, following a stormy debate, was thrown out by the Irish Parliament. Halifax, the Lord Lieutenant, explained that government supporters had taken alarm at the word 'papist' and refused to countenance the proposal which was then promptly dropped.[16] The approaching end of the war – and the end of large-scale recruiting – caused Trimleston's proposal to lose urgency; but though it came to nothing, considerable significance is attached to it. Trimleston's claim that Irish Catholics of property though 'in spirituals ... Roman Catholick' were in 'temporals and in politics ... as much ... Protestant as any ... of his [majesty's] subjects' was noted and his assertion that they wished to help militarily in defending the empire was tacitly accepted.[17] Henceforth the recruitment of Irish Catholics would proceed without fear of legal sanction, though once again not entirely openly. In 1770, Lord Townshend, the Lord Lieutenant of Ireland, claimed that so far as the recruitment of Irish Catholics was concerned, he was quite prepared 'to do the thing' though he felt it prudent to 'avoid the name'.[18] Moreover, around the same time, the enlistment of Irish Catholics took place into the East India Company army.[19] The provision of a special oath of allegiance for Irish Catholics in 1774 was of a piece with these developments. Indeed, it merely put into legal terms what Trimleston had earlier claimed. Elsewhere, too, there were changes: the position of the Catholic church was safeguarded both informally in such West Indian islands as Grenada and Guadaloupe, and formally in Canada in 1774 by the Quebec Act. It was soon claimed that this latter act had a military dimension and that orders had been given to raise a Canadian regiment of French Papists.[20] The conclusion is clear: under the pressure of war and under the strain of garrisoning a new and far-flung empire – one moreover that was religiously divided – British ministers were being forced to look afresh at earlier exclusionist policies. Such a reappraisal could only be to the advantage of Irish Catholics who clearly possessed what the new empire needed: numbers.

Throughout the 1770s as relations between Britain and the American colonies deteriorated, more and more Catholic recruits were taken into the armed forces. Not surprisingly, Irish Protestants grew restive at this development: there were protests at the open recruitment of Catholics into the East India Company army, and alarm was expressed at reports that 10,000 Catholics were to be enlisted for the defence of Ireland.[21] Protestant sympathy with the Americans was increased by the fear that the British government appeared intent on arming Irish, French and American Catholics to coerce them; and there were suspicions that the British government in its eternal quest for troops was not above offering Catholic relief in return for Catholic recruits. Nor were these suspicions groundless. From the late 1770s there was in fact a plan to offer concessions to the Catholics of England, Scotland and, especially, Ireland in order to help recruitment, and at the same time to help fix the loyalty of the various Catholic populations and so enable North's government to present a 'united front' against the Americans.[22]

The scheme involved the Scottish 'projector' and general busybody, Sir John Dalrymple, who was authorised by Lord North to sound out various Catholic leaders in Scotland and in England on a proposal of concessions in return for recruits. The response was encouraging: but given the number of English and Scottish Catholics (about 110,000 out of a combined population of over 7,000,000) North cannot have expected great things from Catholic relief in Britain. It seems clear that Ireland and Irish Catholics was the object aimed at all along. The terms of the English Catholic Relief Act of 1778 were chosen with a strict eye as to what would go down in Ireland: significantly a proposal to alter the oath in the English bill to make it meet English conditions was abandoned lest it cut across the Irish act for which the English one was designed to pave the way. As Edmund Burke put it to a correspondent in Ireland, the English act was 'ultimately intended for you'; and in the debates in the British parliament, the view was expressed that 'this bill ... would when passed, be an example to the Irish parliament.'[23] So it proved. Introduced by Luke Gardiner, who had been on record from an early date as a supporter of Catholic recruitment into the British army, the Relief Bill, closely modelled on the English act was put through the Irish parliament in June 1778 and set the seal on Catholic support for the American war.

When the government stockpurse for recruits ran out, it was the wealthy Catholic merchant, George Goold, who offered to advance the government some 6000 guineas 'and if necessary as much more tomorrow' in order to make up the shortfall; and the Catholic hierarchy proclaimed various fasts for the success of British arms in North America, Bishop Troy looking forward unashamedly to the downfall of the 'Puritan party, calvinistical and republican' in the colonies.[24] More importantly, Irish Catholic recruitment increased while the number of Irish recruits to the French service fell away dramatically so that by the eve of the French Revolution Irishmen made up only from 6 to 10 per cent of the other ranks in the so-called Irish Brigade.[25] Even Catholic officers, though legally barred from the King's army, began, quite improperly, to receive commissions; and in the milder climate that now prevailed, throughout the 1780s various memorials were received from Irish Catholic officers in continental services seeking ways to transfer to the British army.[26] There were many problems still to be overcome, but by the end of the American war 'the idea of Roman Catholic relief as part of wartime strategy' had been firmly established.[27]

Up to and including the War for American Independence, those Irish

Catholics recruited into the British army or the marines had been destined to serve abroad, or aboard ship: and this had been reassuring for those who instinctively doubted the wisdom of instructing any Catholic in the use of arms. But what of Catholics being permitted to participate in the defence of Ireland? And when would the position of Catholic officers be regularised? These were weighty matters which ought to have been settled in the more leisured peacetime atmosphere of the 1780s. This opportunity was not taken: at least it was not taken by either Dublin Castle or Whitehall and it was the Volunteers, the citizen army formed to defend Ireland during the American War, that took the lead at the end of the war in opening many of its corps to Irish Catholics. The fact that a number of Volunteer companies threw open their ranks to Catholics was generally regarded as a sign that volunteering was in decline, that 'the armed property' of the country was yielding to 'the armed beggary'; and Dublin Castle and the London government derived some consolation from this.[28] At the same time, however, the spectacle of Catholics openly bearing arms caused much alarm and led the Lord Lieutenant, Rutland, to urge the outlawing of the Volunteers altogether and the setting up in their place of an Irish Militia, formed on a wholly Protestant basis. It was further evidence of how much Catholic sensitivities were being taken into account by London that Rutland's 'favourite project' was turned down in January 1786. Sydney, the Home Secretary, explained that since 'the militia, it is taken for granted, must be a Protestant one, the Roman Catholics cannot but be offended at being deprived of their arms which will be put into the hands of Protestants. And though it is not advisable to allow the Catholics power, it is extremely unadvisable (*sic*) to give them offence and mortification.'[29] But with the outbreak of the French Revolution and the drift towards a European war in 1792 such sensitivities had to be shelved: military necessity demanded that 'offence and mortification' would have to be given; they would, however, be endured by Irish Protestants.

The resurgence of Volunteer corps in the early 1790s, the pro-French stance of their public utterances, and their call to Irish Catholics to make common cause with them, prompted the Castle's decision to put an end to volunteering and substitute a Militia ('to render individual armies unnecessary') in its place. This Militia,[30] however, was to be entirely different to that proposed by Rutland and, more recently, by the earl of Westmorland, for it was to be balloted indiscriminately from Irish Catholics and Protestants. The property qualifications for officers meant inevitably that a great preponderance of Protestants would have rank; equally, the lack of property qualifications for the men meant, just as inevitably, that the new force would be overwhelmingly Catholic in composition. It was no coincidence that this Militia Act was put through the Irish parliament at the same time as the major Catholic Relief Act of 1793 was reaching the statute books. Nor was it a coincidence that war broke out with France as these acts gained the royal assent. A year earlier more modest concessions for Catholics had been rejected by Dublin Castle but in the winter of 1792–3 the situation on the continent had deteriorated appreciably and war loomed. The demands of war once again forced the pace. Time and time again, Pitt and Dundas made the point that the 'present state of the world', and 'the present circumstances of this country and Europe' made it vital to adopt those measures capable of 'conciliating the Catholics as much as possible and ... making of them an effectual body of support'.[1] The half measures and grudging concessions of 1792 would by no means suffice. Pitt and Dundas wanted to 'connect all lovers of order and good government in a union of resistors to all the abettors of anarchy and misrule', and they, therefore, could see 'no reason why in respect of arms they (Irish Catholics) are to be distinguished from the rest of his majesty's subjects'.[2] Hence, not only were Irish Catholics to be allowed, even forced, into forming a defence force for their country but also, in an equally radical departure from previous practice, there was to be provision for giving commissions in the British army to those of them who had the required qualifications.

As a result of the Militia Act and the Catholic Relief Act of 1793, Irish Catholics now had the vote on the same terms as Irish Protestants and clearly they were expected to play a major part in the defence of Ireland during the war with revolutionary France. Admittedly, those Catholics appointed officers would only have a valid commission while they served in Ireland for the act did not extend to England: but none the less it was incontestable that these changes amounted to both a military and a political revolution. Within a generation, the British state had gone from a policy of firm exclusion of Catholic soldiers to one of forced inclusion; from fear of Catholic numbers to reliance on them. Not surprisingly, it was now Irish Protestants who felt betrayed and not a little bewildered at the speed of this revolution, and it was with much grumbling and resentment that the Catholic bill was put through the Irish Parliament. Moreover, within a year there was a further startling reversal of policy when the Irish Brigade of France, long-standing bugaboo of Irish Protestants, was taken onto the Irish establishment. This experiment ultimately proved a failure but it has to be seen as part of that reversal of policies, that overturning of prejudices, that under the pressure of war constituted the revolution of 1793.[3]

With a largely Catholic Militia in the front line of the defence of Ireland,

and with huge numbers of Irish Catholics being recruited into the regiments of the regular army (and into the Royal Navy), it seemed inconceivable that the major remaining element in what had come to be known as Catholic emancipation – the right of Catholics, if elected, to take their seats in the Irish parliament – would be long delayed. Catholic expectations in this respect seemed to be fulfilled with the appointment of the Whig Earl Fitzwilliam as Lord Lieutenant. Not only was Fitzwilliam on record as being in favour of Catholic emancipation but he was also convinced that the successful prosecution of the war against revolutionary France demanded it. He believed that his predecessor, Westmorland, had shamefully neglected the defences of Ireland and it was a central element of Fitzwilliam's mission (the word is not too strong) in Ireland 'to make the experiment of the ... loyalty and zeal of the Catholics'.[34] By this he meant the establishment of new forces which would further tap into the Catholic manpower reservoir.

Initially, Fitzwilliam floated the idea of a Catholic Brigade drawn from Irish Catholics, and he even contemplated the formation of an Irish Presbyterian Brigade.[35] His main thrust, however, was to set up a Yeomanry – 'an armed constabulary composed of the better orders of people' – and as such a force would necessarily 'be compos'd of a certain description of the tenantry, a very large proportion of whom are Catholic', Catholic emancipation had to be conceded. The establishment of a Yeomanry, he wrote, 'must not precede the Roman Catholic business ... for should the Catholic question fail we must think twice before we put arms into the hands of men newly irritated.'[36] Once again we are back in the familiar realm of Catholic relief as a part of wartime strategy: this time, however, the equation was not made. Fitzwilliam might argue that Catholic emancipation would 'rivet [Catholic leaders'] affections and stimulate them to exercise whatever influence they possess over the great mass of the people' and thus enable a Yeomanry to be embodied:[37] but London was not convinced that emancipation was either necessary or desirable, nor indeed that a Yeomanry was at that time practicable. In its eyes the concession account for Irish Catholics had been closed in 1793 and, as violence mounted in Ireland and as success in the war continued to prove elusive, the British government looked more and more to the Irish governing élite to provide the necessary leadership to see Ireland through the crisis. The price of that leadership was Protestant Ascendancy. Catholic emancipation as proposed by Fitzwilliam would not only raise the spectre of parliamentary reform and perhaps endanger the established church in Ireland: it would also destroy Protestant confidence. Ironically, it was now claimed that a period of war was no time for experiments.

Fitzwilliam was recalled and his policy abandoned by the Pitt cabinet. But the establishment of a Yeomanry remained on the agenda and it was duly embodied the following year. The force was, however, far removed from that which Fitzwilliam had envisaged for, as constituted, it amounted to little more than, as John Beresford put it, 'arming the Protestants who can be depended on'.[8] Catholics were by no means excluded, but from 1797 on the trend was unmistakably to protestantise the force and thus to erect a standard to which loyalists would flock. Fitzwilliam's ideal, that the 'people might be made one people, one Christian people' and that the Yeomanry might reflect this union, was nowhere to be seen.[39] His ideal had not, however, entirely disappeared, for William Windham, British Secretary at War in the 1790s and again in the Grenville All-the-Talents ministry of 1806–7, never lost sight of it. In fact, Windham's and Grenville's pursuit of it was ultimately to bring down their administration in 1807.

The war with France had continued to go badly throughout the 1790s, and Windham, charged with prosecuting it, had grown more and more convinced that extraordinary measures would be needed before victory could be assured. Taking a leaf out of the French book, he began to call for the creation of 'an armed nation' in both Britain and Ireland which could more effectively confront French expansionism. So far as Ireland was concerned, he wanted 'the property of that country to arm themselves'; then 'the most rigourous and prompt measures' could be employed against subversives. Such measures, added Windham, 'must not be unaccompanied by others' and he suggested 'a Catholic establishment for the Catholics and an exemption from tithes'.[40] Nothing came of this idea at the time. Within a month, the rebellion had broken out, exposing the bankruptcy of both the Irish parliament and Dublin Castle, and at the same time offering Pitt the opportunity of putting through a union between Ireland and England. This union was primarily legislative; but it was also a military union and strategic considerations lay behind it. It integrated the Irish armed forces, particularly the Irish Militia, more fully into the war effort and in theory ought to have brought about that 'armed nation' that Windham had called for. The failure to put through Catholic emancipation, or indeed any concession, however, endangered this strategy. In a modest attempt to remedy this failure, Grenville and Windham on attaining office in 1806, took up the issue of the legal standing of Catholic officers serving outside Ireland and sought to have their rights safeguarded; they also tried to address the problem of Catholic soldiers being forced to attend Protestant worship. The resultant uproar brought down the government; but it also illustrated graphically the changed nature of the connection between the

Catholic question and the demands of war.

In the 'revolution of 1793' Irish Catholics balloted into the Militia had been entrusted with a major role in the defence of Ireland against both invasion and insurgency. In general, the Irish Militia had fulfilled this role satisfactorily and its record in the field had been a fairly creditable one. Its reputed disaffection, however, allied to its notorious indiscipline and its disgrace at Castlebar in 1798 had aroused misgivings about the reliability of the force.[41] But more influential than anything, as government in Ireland in the later 1790s committed itself more and more to support of the Ascendancy, the Irish Militia's overwhelming Catholic composition had forced a reappraisal of its role. As the Catholic defenders of a Protestant state, the Irish Militia appeared an anomaly, an embarrassment, perhaps even a threat. Greater reliance was now placed on the largely Protestant Yeomanry and it was the Yeomanry, not the Militia, which figured in the defence plans drawn up in the aftermath of the rebellion.[42] The Militia were dismissed as worthless: 'I need not tell you' Cornwallis informed Dundas, 'how little dependence is to be placed on the Irish Militia serving in their own country';[43] and accordingly a scheme was soon underway to send them abroad. In 1800, Cornwallis offered four guineas to each militiaman who would volunteer for regular service in a line regiment; and he offered an ensigncy to each Militia colonel who enlisted forty volunteers from his regiment. By 1805 a bounty of between ten and twelve guineas, depending on the terms of service was being offered and about 3000 militiamen per annum (some years many more) were passing into the line regiments for service abroad.[44] From being in the vanguard of the defence of Ireland, the largely Catholic Militia had become merely a 'nursery for the regulars', while at the same time the largely Protestant Yeomanry continued to expand and take on new duties.[45] That commitment to Protestant principles which Edward Cooke, the former under-secretary in Dublin Castle, claimed distinguished post-union government in Ireland was thus given the most explicit and tangible form.[46]

So many Irish Catholic soldiers were recruited into the British army for service overseas and so many Irish Catholic officers were now serving in Great Britain that the military authorities faced a number of problems. By the Catholic Relief Act of 1793, Catholics could obtain commissions in the Irish army, and Militia; and there had been a general understanding that the religious sensitivities of Irish Catholic recruits would be respected while regiments remained in Ireland. But no such legal basis, and no such understanding existed in England: Catholic officers in theory had to give up their commissions when they transferred to England; and Catholic soldiers frequently found themselves compelled to attend Protestant worship when in England. Quite how genuine these grievances were is a debatable point. In 1798, Lord Chief Justice Mansfield's judgement in the Petrie case, which concerned the legality of a commission for an English Catholic, stated that Catholics could take a commission but were liable to swear the oath of supremacy *if called upon to do so*. Scott and Mitford, respectively Attorney and Solicitor-Generals concurred: commissions for Catholics in the British army may have been repugnant to the constitution but they were improper rather than illegal.[47] This appears to have been the case in practice for a long time. As early as 1780 the Catholic Archbishop of Dublin had sought guidance from Rome as to the morality of a Catholic concealing his religion in order to receive a commission; and clandestine commissions must have been relatively common in the 1780s.[48] Nearly thirty years later, Sir Arthur Wellesley declared that 'it was notorious that no officer of the army and navy had been required for many years to take any oath or to qualify in any manner; and that it was equally notorious that there were many Roman Catholic officers in the King's service.'[49] Even Grenville and his brother Buckingham conceded that 'Irish Catholics ... find no difficulty in getting commissions in the army' and that 'there are many [Catholic officers] in our army'.[50] None the less, there were still difficulties: in 1803 Captain George Bryan wrote to Charles James Fox complaining that once his religion – he was a Catholic – was discovered, the King was furious that he had got a commission and he was now under pressure to go on to the half-pay list.[51] Ex-Lieutenant Henry Hamilton of the North Down Militia claimed that he was kept out of the army because he was a Catholic. There may have been other causes at work here: in his letter to the Prince of Wales protesting his lot, he was imprudent enough to remind the Prince of the fate of Spencer Perceval, murdered by a madman whose petition he had ignored.[52] These were isolated cases: in the vast majority of instances Catholic officers incurred no penalty.[53] In any case, whether the grievance was a practical one hardly mattered: to Grenville's tidy mind, connivance and blind-eye turning were hardly acceptable ways to order the armed forces of the crown and he was determined to eliminate the ambiguity, 'the manifest incongruity', of this point.[54]

There was also a wider issue at stake here. In theory access by Catholics to many of the higher offices in Ireland had been opened since 1793 but, ten years on, their representation in those offices had not increased dramatically. Without doubt the Catholic middle, or propertied, classes resented this exclusion from possession of the loaves and fishes of political and military careers. Lord Harwicke, the first Lord Lieutenant after the Union, explained to Lord Hawkesbury why he understood the Catholic middle classes were impatient for relief. One cause 'lay principally to the increased wealth and consequence which many of the Roman Catholics have derived from trade and to the desire they feel of not being excluded by their religion from those objects of pursuit which are open to the rest of their countrymen'. Hardwicke went on:

Formerly the Irish Brigade in France, the Spanish and particularly the Austrian Service opened a wide field for the younger sons of Roman Catholic families; that in the year 1779 there were upwards of 600 Irish officers in the Austrian army; that at present there is scarcely one third of that number and of those many are not Irishmen by birth. That notwithstanding the Irish Act of 1793 which permits Roman Catholics to have commissions in the army there are very few in his majesty's service in which they cannot be employed in any higher rank than that of colonel.[55]

Grenville's plan aimed at remedying this situation.

Secondly, Grenville was convinced that the true potential of Ireland's manpower reservoir had not yet been fully exploited. Within a short time of attaining office he was pressing the Duke of Bedford, Lord Lieutenant of Ireland, and William Elliot, the Chief Secretary, to investigate the possibility of recruiting Catholic regiments for service overseas in places like Sicily, Malta, Portugal or even South America. 'We want the men', wrote Grenville, 'Ireland wants a vent for its superabundant population: could not these two wants be reconciled?' In order to encourage Catholic recruiting, he suggested the appointment of Catholic chaplains to these regiments; and there would be a public declaration that Catholics would have the free exercise of their religion when serving in the army. Grenville also proposed an additional clause to the annual Mutiny Act permitting the King 'whenever he shall deem it advisable to confer any military commission whatever on any of this liege subjects'.[56] This last inducement was to prove his undoing. George III had all along believed that what was being asked of him was his agreement to the extension of the Irish 1793 Act so far as it related to army officers only; he soon realised that Grenville's proposals were much more far-reaching than that for they embraced all ranks in the army, navy and marine. Claiming that he had been misled, the King refused to countenance any further indulgences. Within a week Grenville's ministry of All-the-Talents had fallen. It was not until 1817 that the position of English Catholic officers in the army and navy was regularised by the simple expedient of grouping them with other Dissenters for the purpose of benefiting from the Annual Indemnity Act.[57]

In many respects, the Grenville débâcle represented the last echoes of that policy which Pitt and Dundas had embarked upon some thirteen years earlier. Fitzwilliam had attempted to take it further in 1795 but he had failed, and though Windham had subsequently endorsed it and much was heard of it at the time of the Act of Union, in the end nothing had been done and it had been left to Grenville in 1806 to take up the question once again. In every case, there was a belief that Ireland and Irish resources – especially human resources – were vital to the war effort, that these resources had been neglected, and that only by uniting the contending peoples of these islands into one 'armed nation' could the war be fought to victory. No matter that tens of thousands of Irishmen were flocking to the colours: it was an article of faith that there were tens of thousands more who hung back. In this endeavour, it was essential to enlist the aid of the priests and the lay Catholic leaders: hence, safeguards for the practice of the men's religion were required and openings had to be given to the Catholic propertied classes. So Fitzwilliam and Grenville and even Pitt at one time had argued: but it was not to be. The hostility of George III to any re-opening of the Catholic question, the desire on many sides to preserve the Protestant constitution and especially the fact that thousands of Irish Catholics were enlisting anyway all combined to render less desirable, less pressing the old idea of concessions in return for recruits. However, there was a danger here, for while the admission of Catholics to the armed forces of the crown had been given as a boon their continued exclusion from full citizenship would be taken as a grievance. Ironically, it could be argued that the war which saw Irish Catholics demonstrating their loyalty in the most positive manner by enlisting in huge numbers – a figure of 200,000 was cited – ended with Catholics more alienated and disaffected than ever.

The huge numbers of recruits from Ireland showed no sign of diminishing in the years after the fall of the Talents, and with the major military commitment on the Iberian peninsular after 1808 even more recruits were sought. Irish Catholics may at one time have constituted a 'weapon of war yet untried', but from an early date this was very far from being the case. Yet this weapon, with the collapse of the Talents and the continued denial of Catholic emancipation, showed some signs of being a double-edged sword. Catholic numbers in the ranks may have demonstrated Catholic loyalty, but they also constituted a powerful argument in favour of Emancipation; and the alleged grievances of Catholic soldiers serving abroad presented the various Catholic Committees and Boards with a powerful issue that united both lay and clerical behind it.

From the time of the Peninsular war on, the size of the Irish contingent in the British army became a routine element in the rhetoric of those seeking Catholic emancipation. As one Dublin Castle official commented: 'The minds of all the parties in the Catholic body are directed towards the events in Spain and whatever be the result, they ultimately anticipate the measure of emancipation'.⁵⁸ Daniel O'Connell himself made frequent reference to the Irish soldiers, sailors and marines who had fought earlier at Aboukir Bay, the Nile, and Trafalgar, and he was not slow to cite those who had recently been in combat at Vimiero and Talavera. Their bravery, he claimed, entitled them and their co-religionists at home to reward.⁵⁹ Similarly, the grievances of those Catholic soldiers also received much attention. There were frequent complaints that colonels regularly ignored the circular orders allowing the men to decide themselves which service (if any) they attended and that Catholic soldiers were forced to attend Protestant worship when serving in England or abroad. 'Every Catholic soldier must hesitate', warned O'Connell, 'before he ventures to a country where the rights of his religion may be denied him'; and in 1811 Bishop O'Donnell of Derry claimed that though the Irish soldiers – he was addressing Militiamen in his congregation – fought bravely, they were not so rewarded as Protestants 'but were passed over and neglected, and none of the laurels of victory crowned the brow of the Roman Catholic.'⁶⁰ Moreover, the Catholic Board itself began to look on the Irish Militia as uniquely its army and to call on soldiers to sign the Catholic petitions.⁶¹

This attempt at politicising of Irish soldiers was deeply worrying for Dublin Castle: the spectacle of the Irish Militia playing a role akin to that of the Volunteers of 1782 or even acting as a sort of military wing of the Catholic movement was particularly disturbing. As a solution to the problem, in 1811 the Militia Interchange Act was enacted over the protests of O'Connell and others who feared that their party would be weakened by it. The act authorised the removal of up to one third of the Irish Militia (about 10,000 men) on an annual basis to serve in Great Britain, and their replacement with Scottish and English Militia regiments. It was anticipated that most of those going to England would ultimately end up overseas. Dublin Castle was only too delighted to get rid of them – 'all the regiments we send you are Catholic except one' wrote Pole, the Chief Secretary – and he urged the cabinet in London to 'take as many of the Irish Militia as you can out of Ireland'.⁶² With the removal of the Irish Militia as a political weapon and, even more important, with the triumphant progress of the war, there was less urgency than ever to do anything about the Catholic question. Lord Redesdale remarked that Napoleon's ill-fated campaign in Russia 'was the most formidable foe to Catholic emancipation', and the point was frequently made that victory in Europe would mean a set-back for the Catholic question. 'The state of the peace of this country', explained Chief Secretary Vansittart in 1814 referring to the Catholic question, 'as long as the oldest man living has known it, has always depended upon the state of the wars in which England was engaged'.[63] Admittedly, both the British Catholic Board and the Irish Catholic Hierarchy claimed emancipation as a reward for the part that the Catholics had played in England's triumph – even Sidmouth conceded that 'the supply of troops from Ireland turn'd the scale on the 18th June at Waterloo' – but their pleas were ignored; and the failure to fulfil the Union understanding was compounded by the failure to reward the Irish Catholics for their part in the victory over Napoleon.[64] A few lines from a ballad that circulated some years after the war made the point well:

> Oh Wellington sure you know it is true
> In blood we were drenched at famous Waterloo
> We fought for our king to uphold his crown
> Our only reward was – Papists lie down.[65]

It might be assumed from the above that the connection between military matters and the Catholic question concluded with the end of the French wars, but this is not so. Elsewhere, I have argued that the experience of those tens of thousands of Catholic soldiers who served in the armed forces during this period had a politicising effect on them which they carried over into civilian life.[66] Sometimes those ex-soldiers made use of their military training when they left the army: there were reports in 1821 that 'disbanded soldiers' were involved in agrarian disturbances, and Grant, the Chief Secretary, was sufficiently impressed by these to have the pensioners in the area concerned take the oath of allegiance.[67] Moreover, soldiering remained a vent for Ireland's surplus population so much so that by 1830 over 40 per cent of NCOs and below of the British army was composed of Irishmen.[68] These Irish soldiers, well represented even in 'English' regiments, had a passive but none the less vital role to play in the agitation that convulsed Ireland on the eve of the winning of Catholic Emancipation. The Catholic Association made much of the point that the Dublin authorities had no choice but to rely on Catholic soldiers to enforce law and order and in an intimidatory way it began to publish the figures for the number of Catholics in various regiments.[69] In another worrying development, Dublin Castle received intelligence reports that the troops might not prove reliable in the event of a crisis. Peel was certainly impressed by this factor and it played a part in his and Wellington's decision to yield. In the event, Catholic Emancipation was conceded in a time of peace, though not one of internal tranquillity; the 'imperative necessity' which moved Peel was not the threat of a foreign invasion, but rather the very real prospect of a civil war.[70] In 1829, as in 1759, war and the Catholic question were intimately connected.

NOTES

1   For a general survey of the Catholic question and Anglo-Irish politics, see Thomas Bartlett, *The Fall and Rise of the Irish Nation: the Catholic Question, 1690–1830* (Dublin 1992).

2   For fuller discussion, see Thomas Bartlett 'The Origins and Progress of the Catholic Question in Ireland, 1690 to 1800', in Thomas Power and Kevin Whelan (eds.), *Endurance and Emergence: Catholics in Ireland in the Eighteenth Century* (Dublin 1990), pp. 1–19.

3   On Catholic wealth, see Maureen Wall, 'The Rise of a Catholic Middle Class in Eighteenth-Century Ireland' and 'Catholics in Economic Life' in Gerard O'Brien (ed.), *Catholic Ireland in the Eighteenth Century: Collected Essays of Maureen Wall* (Dublin 1989), pp. 73–84, 85–92.

4   For some scepticism on this point, see R.B. McDowell, *Ireland in the Age of Imperialism and Revolution* (Oxford 1979), pp. 179–80.

5   See Colin Haydon, 'Anti-Catholicism in Eighteenth-Century England' (unpublished D.Phil. thesis, Oxford University 1985).

6   See R. Kent Donovan 'The Military Origins of the Roman Catholic Relief Programme of 1778', *Historical Journal*, xxviii (1985), pp. 79–102; see also R.E. Burns, 'The Catholic Relief Act in Ireland, 1778', *Church History*, xxxii (1963), pp. 181–206.

7   Donovan argues this point in relation to the 1778 Relief Act in his article 'Military origins'.

8   Maureen Wall, 'The Penal laws', reprinted in O'Brien (ed.), *Catholic Ireland*, pp. 16, 19–20.

9   P.G.M. Dickson, *The Financial Revolution* (London 1967); John Brewer, *The Sinews of Power* (Yale 1989).

10   See Geoffrey Best, *War and Society in Revolutionary Europe* (London 1982), *passim*; Clive Emsley, *British Society and the French Wars, 1793–1815* (London 1979), pp. 33, 94, 133; for recruitment during the Seven Years' War, see C.R. Middleton, 'The Recruitment of the British Army, 1755–1762', *Journal of the Society for Army Historical Research*, lxvii, no. 272 (Winter 1989), pp. 226–38.

11   Donovan, 'Military Origins', p. 93.

12   Thomas Bartlett, 'Army and Society in Eighteenth-Century Ireland', in W.A. Maguire (ed.), *Kings in Conflict: the Revolutionary War in Ireland and its Aftermath* (Belfast 1990), pp. 173–85; Ira D. Gruber, 'On the Road to Poonamalle: an Irish Officer's View of the War for American Independence', *The American Magazine and Historical Chronical*, 4, no. 1 (Spring-Summer 1988), pp. 1–12. I am indebted to Mrs Arlene Shy for calling my attention to this article.

13   Bruce Lenman and John Gibson (eds), *The Jacobite Threat: Rebellion and Conspiracy: A Source Book* (Edinburgh 1990), pp. 238–9; Sir Henry McAnally, 'The Militia Array of 1756 in Ireland', *Irish Sword*, i (1949–51), pp. 94–104.

14   British regiments serving in Canada in the Seven Years' War were described as being composed of 'convicts and Irish papists': William Fitzwilliam to Lord Fitzwilliam, 27 Sept. 1755 (Public Record Office of Northern Ireland [PRONI], Pembroke Estate Papers).

15   R.E. Burns, *Irish Parliamentary Politics in the Eighteenth Century*, vol. ii, 1730–60 (Washington DC 1990), pp. 278–9.

16   Halifax to Egremont, 13 Apr. 1762 in *Calendar of Home Office Papers, 1760–75* (4 vols, London 1878–99), i, p. 173.

17   Trimleston to Egremont, 5 Mar. 1762 (W. Sussex Record Office, Petworth House Mss1270).

18 Note by Townshend, c.27 Dec. 1770 (PRO Chatham Mss30/8/84/103).

19 Clive was a strong advocate of ignoring the religious bar. See quotation in Joel Mokyr and Cormac Ó Gráda, 'The Height of Irishmen and Englishmen in the 1770s: Some Evidence from the East India Company Army Records', *Eighteenth-Century Ireland*, iv (1989), pp. 83–4.

20 David Milobar, 'Conservative Ideology, Conservative Government and the Reform of Quebec, 1782–1791', *International History Review*, xii, no. 1 (Feb. 1990), pp. 45–64; Donovan, 'Military Origins', p. 83; J.R. Hill, 'Religious Toleration and the Relaxation of the Penal Laws: The Imperial Perspective, 1763–1780', *Archivium Hibernicum*, xlv (1989), pp. 98–109.

21 Mokyr and Ó Gráda, 'Height of Irishmen', p. 85; Lord Shannon to Wm. Dennis, 17 Aug. 1777 (PRONI, Shannon Mss D2707/A/2/3/45).

22 Donovan, 'Military Origins', *passim*.

23 Nigel Abercrombie 'The First Relief Act', in Eamon Duffy (ed.), *Challoner and his Church* (London 1978), pp. 189–90.

24 George Goold to General Irvine, 13 Sept. 1771 (PRO SP63/476/141); Troy to the Papal Nuncio, 27 Oct. 1777 (Dublin Diocesan Archives [DDA], Troy Mss29/6/40).

25 See Samuel Scott, 'The French Revolution and the Irish Regiments in France', in David Dickson and Hugh Gough (eds), *Ireland and the French Revolution* (Dublin 1990), p. 15.

26 See the example in The Macarthy Mor, '"By the Help of Many Lies": How Penal Were the Penal Laws?', *Familia*, ii, no. 6 (1990), p. 59.

27 Donovan's phrase, 'Military Origins', p. 102.

28 See James Kelly, 'A Secret Return of the Volunteers of Ireland in 1784', *Irish Historical Studies*, xxvi, no.103, pp. 268–92.

29 Sydney to Rutland, 7 Jan. 1786 (His. Mss Comm. *Rutland*, iii, p. 273).

30 See Thomas Bartlett, 'An End to Moral Economy: The Irish Militia Disturbances of 1793', *Past and Present*, no. 99 (May 1983), pp. 41–64; and Sir Henry McAnally, *The Irish Militia: A Social and Military Study* (Dublin 1949), pp. 1–42 for the origins of the 1793 Act and the reaction to it.

31 Pitt to Westmorland, 10 Nov. 1792 (PRO HO100/38/373–4); Dundas to Westmorland, 17 Dec. 1792 (PRO HO100/38/157–60).

32 Dundas to Westmorland, 7 Jan. 1793 (PRO HO100/43/128–43).

33 For material on the Irish Brigade, see National Archives, Dublin, O[fficial] P[apers] 23/8/1–5. Ultimately, after many set-backs, the six battalions were merged into two regiments, both of which served continuously in North America.

34 Fitzwilliam to Portland, 15 Jan. 1795 (PRO HO100/46/264–51).

35 Ibid.

36 Fitzwilliam to Portland, 10 and 28 Jan. 1795 (PRO HO100/56/57–9 and 46/268–9).

37 Fitzwilliam to Portland, 15 Jan. 1795 (PRO HO100/56/229).

38 Beresford to Auckland, 4 Sept. 1796, in Wm. Beresford (ed.), *Correspondence of the Rt. Hon. John Beresford* (2 vols, London 1854), p. 129.

39 Fitzwilliam to Portland, 10 Feb. 1795 and enclosed 'Proposed Oath for Catholics' (PRO HO100/56/222–8, 229).

40 Wm. Windham to Dundas, 8 Apr. 1798 (W.L. Clements Library, Ann Arbor, Michigan, Melville Papers).

41 For a discussion of these charges, see Thomas Bartlett, 'Indiscipline and Disaffection in the Armed Forces in Ireland', in P.J. Corish (ed.), *Radicals, Rebels and Establishments* (Belfast 1985), pp. 115–34.

42 See for example Wellesley to Hawkesbury, 7 May 1807 in Duke of Wellington (ed.), *Civil Correspondence and Memoranda of The Duke of Wellington: Ireland, 30 Mar. 1807 to 12 Apr. 1809* (London 1860), v, pp. 28–36.

43 Cornwallis to Dundas, 19 Jan. 1799 in Chas. Ross (ed.), *Correspondence of Cornwallis* (3 vols, London 1859), iii, pp. 117–8,

44 Cornwallis to Portland, 31 Jan. 1800 (PRO HO100/90/31–2); 'Proposal for Recruiting the Militia', 25 Mar. 1805 (PRO HO100/125/221–3).

45 Col. T.H. Foster to John Foster, 4 Apr. 1804 (PRONI, Foster Mss D207/11/46).

46 Ed Cooke to Camden, 13 Sept. 1803 (PRONI, Camden Transcripts T2627/4/136).

47 Note on the Petrie case, June 1798 (n.d. *c.*June 1806) (B.L. Add. Mss47569/301–2).

48 (Archbp Carpenter) to Congregation of the Holy Office, *c.*1780 (DDA, 116/3/13).

49 Wellesley's Memorandum, 1807 in Wellington (ed.) *Civil Corr.*, v, pp. 14–15.

50 Grenville to Bedford (Dec. 1806) in Hist. Mss. Comm. *Dropmore*, viii, pp. 486–8; Buckingham to Grenville, 11 Feb. 1807, Hist. Mss. Comm. *Dropmore*, ix, p. 65.

51 Capt. George Bryan to C.J. Fox, 18 Dec. 1813 (B.L. Add. Mss. 47569/149).

52 Henry Hamilton to Prince of Wales, 9 Dec. 1803 (PRO HO100/175/120–3).

53 This statement is made with some caution: there may have been problems for Catholic officers in the other services. Capt. Edward Whyte claimed that he had served from 1796 to 1814 in the Royal Navy, but that recently he had been told to take the oaths and on his refusal had been drummed out of the navy and his half-pay stopped (Case of Edw. Whyte, *c.*1816 Nat. Lib. Ire., Grey Mss., Pos. 3709).

54 Grenville to Bedford, (Dec. 1806) in Hist. Mss. Comm. *Dropmore*, viii, pp. 486–8.

55 Hardwicke to Hawkesbury, 26 Nov. 1804 (PRO HO100/123/188–91).

56 Grenville to Elliot, 1 Aug. 1806; Grenville to Bedford, (Dec.) 1806: Hist. Mss. Comm. *Dropmore*, viii, pp. 253, 491–4.

57 There had in fact been some talk of extending this Act to England in 1812 – even Sidmouth, notoriously anti-Catholic, admitted that 'consistency and justice' required it, but nothing was done, probably lest the Catholic question was raised once again. When the matter was resolved, as it was in 1817, it was done 'slyly', without much publicity. Sidmouth to Richmond, 19 Nov. 1812 (Nat. Lib. Ire., Richmond Papers, 65/823); Edw. Hay to Donoughmore, 5 Dec. 1817 (PRONI, Donoughmore Transcripts, T3459/D/16/18).

58 E.B. Littlehales to …, 27 Nov. 1808 (PRO HO100/149/227).

59 Speech by O'Connell, 2 Feb. 1811, in John O'Connell (ed.), *The Life and Speeches of Daniel O'Connell* (2 vols, Dublin 1846), i, p. 78.

60 Speech by O'Connell, 28 May 1811, ibid., pp. 107–8; Thomas Bartlett, 'Militarization and Politicization in Ireland, 1780–1830' in L.M. Cullen and Paul Bergeron (eds), *Culture et pratiques politiques en France et en Irlande, xvie-xviiie siècles* (Paris 1991), pp. 125–36, quotation at p. 132.

61 Pole to Ryder, 27 May 1811 (PRO HO100/163/319); McAnally, *Irish Militia*, p. 245; Col. Akenson to Littlehales, 19 Jan. 1810 (PRO HO100/158/82).

62 Pole to Ryder, 23, 26 July 1811 (PRO HO100/164/104, 147–9).

63 Redesdale to …, 24 Sept. 1813 (PRO HO100/172/86–9); Vansittart to Sidmouth, 19 Jan. 1814 (PRO HO100/176/421–4).

64 British Catholic Board, 31 Jan. 1816 (Nat. Lib. Ire. Grey Mss. Pos.3709); R.C. Prelates to Prince Regent, Aug. 1815 (PRO HO100/185/12); Sidmouth to Whitworth, 24 June 1815 (PRO HO100/184/204).

65 Ballad enclosed in Warburton to Gregory, 18 June 1829 (BL Add. Mss. Peel-Gregory

Papers, 40334/299).

66   Bartlett, 'Militarization and Politicization'.

67   Grant to Hobhouse, 28 Nov. 1821 (PRO HO100/202/211–13).

68   H.J. Hanham 'Religion and Nationality in the Mid-Victorian Army', in M.R.D. Foot (ed.), *War and Society* (London 1973), pp. 176–81.

69   John Reynolds, *The Catholic Emancipation Crisis in Ireland, 1823–29* (New Haven 1954), p. 147n.

70   See Peel's speech of 5 Mar. 1829 in *The Speeches of the Late Rt. Hon. Sir Robt. Peel, Delivered in the House of Commons* (4 vol. reprint edn, New York 1972), i, p. 699.

Irene Collins

Napoleon's parliaments, unlike his wars, are not a well-known feature of his *régime*. Georges Lefèbvre, in a monumental study of the Empire, devoted fewer than 100 lines in all to the parliaments, scattered over the 600 pages of the volume.¹ This meagre record, which was about par for the course, has doubtless been a source of disappointment to Napoleon's shade, in whatever Valhalla it may be resting; for he regarded himself as an expert on parliamentary matters, and was proud of the part he had played in establishing a system which he believed was suited to the French character. Above all, he was proud of the legislation which had been achieved by means of the system, and which he believed to be as worthy of fame as his victories on the battlefield. The combination of classical and military education which Napoleon received at Brienne had taught him that lasting glory can be achieved in this world either by fighting great battles or by framing great laws. The most renowned of classical heroes had attempted both, thereby establishing the tradition that the two were somehow connected: Napoleon's particular idol was Lycurgus, the semi-mythical law-giver believed by the ancient world to have been commissioned by the Delphic oracle to reform the constitution of Sparta along military lines. Napoleon was eager to be included in the same category as such men, and he was never more pleased than when Jacques-Louis David, at the request of a Scottish admirer, painted a portrait which showed him standing in his study with a copy of the Code Napoléon visible in the glass-topped desk beside him and a map of Austerlitz spread out on the floor.²

Since a great deal of attention has been paid to the strategy by which Napoleon won his resounding victories, and very little to the methods by which he produced the codes of law that were to spread across the civilised world, a brief introduction to the latter may be in order.³ Napoleon, like Rousseau, made a distinction between laws which were of transitory significance and could be effected by decree, and fundamental laws destined to regulate people's lives for generations to come. The latter, though they might also be framed by one man or by a committee of experts – Napoleon took responsibility for producing them himself with the help of a hand-picked Council of State – should be submitted to the people for acknowledgement as emanating from their will. The Jacobins, in their still-born constitution of 1793, had envisaged putting all such laws to the people in the form of a plebiscite. Napoleon adopted plebiscites for constitutional changes – in 1799, 1802, 1804, and even during the Hundred Days – but the rest of his laws he laid before a *Corps Législatif*, a gathering of deputies from every department of France, whose task was to say yes or no to them as the people would have done in a referendum. They were not allowed to discuss them; they were merely allowed to take a certain amount of advice. In the early years there was an elected body called the Tribunate which studied bills and made a report of them to the *Corps Législatif*; later, when Napoleon had grown tired of criticism from the Tribunate, committees of the *Corps Législatif* itself were allowed to study bills and submit a report to the rest of the body. The legislators maintained complete silence throughout this process, and voted yes or no by placing white or black discs in an urn.

Napoleon's parliaments thus played no part in the formation of laws; yet they were offered a large share in the glory accruing from the finished product. Pomp and ceremony surrounded the opening and closing of the sessions, and Napoleon's spokesmen were lavish in fulsome flatteries. Every attempt was made to associate the achievements of parliament with those of the armies on the battlefield, for were not the one laying the foundations of freedom at home whilst the other staved off threats from despots abroad? 'Gentlemen, Deputies from the Departments to the *Corps Législatif*, I count on your assistance as I do upon the bravery of my army', Napoleon told the legislators in February 1805, as his army lay encamped at Boulogne for the invasion of England.⁴ Once the fighting began, he continually tried to symbolise the connection by sending parliament flags he had seized in battle – sometimes as many as 70 at once.⁵ Phillippe-Paul de Ségur, whose description of the Moscow campaign is an acknowledged masterpiece of flamboyant literature, recounted in equally impassioned terms his experiences as a young colonel, commissioned to carry flags from Madrid in 1808 and present them to the *Corps Législatif*. He was escorted by a company of infantry, which bivouacked around him at night whilst he made himself as comfortable as he could in a carriage stuffed with conquered flags. On arrival in France, the jolting of the carriage re-opened a wound he had received in the Battle of Burgos, and as he could not possibly entrust his sacred mission to anyone else the presentation had to be put off to the next session. When the ceremony finally took place, he was so overcome by what he described as 'the honour of speaking before the representatives of the greatest of nations, in the name of its Grand Army and of the greatest of men' that he was rendered tongue-tied, until his faculties were restored by the need to give a familiar word of command to an officer who had taken a false step beside him.[6]

In 1806 Napoleon decided to cast his message in tablets of stone. The Palais Bourbon, where the deputies sat, he had always regarded as a mean building, standing on low ground with its principal entrance turned away from the river. It was now to be given an imposing classical elevation on the river front, with 30 steps leading up to a portico of twelve Corinthian columns surmounted by a sculptured pediment depicting the Emperor presenting flags to the *Corps Législatif* after the Battle of Austerlitz.[7] This, he said, would give the building an appearance suitable to a Temple of Laws, and make it into a fitting complement to the Temple of Glory he was in the process of creating on the opposite bank of the Seine (the church of the Madeleine, which had been under construction since before the Revolution, had been taken over by Napoleon as a suitable setting for memorials of his triumphs). To his sorrow, neither project was a complete success. The pediment over the Temple of Laws was an embarrassment as soon as it was finished, because France was by then allied with Russia, and Napoleon had to order the police to discourage snide remarks about the inglorious behaviour of Tsar Alexander on the field of Austerlitz.[8] At the other side of the river the completion of the Madeleine proved to be extremely costly, and was not yet finished when Napoleon made a tour of inspection in 1811. By that time the difficulties of the Spanish campaign were causing him to suspect that Fortune, the goddess whom he had so long worshipped, had turned against him, and that he had better give back the Madeleine to the church for the worship of God.[9] Meanwhile there is no direct evidence as to what the legislators thought about such grandiose concepts of their role, but it was probably not much. They were, after all, mostly civilian: of the 11,000 men known to have sat in Napoleon's parliaments at one time or another, it is unlikely that more than 10 per cent had seen military service.[10] They treated the pseudo-military uniform they were supposed to wear with the disrespect of naughty schoolboys: police agents in the public galleries continually reported large numbers as improperly dressed,[11] and according to an English visitor to one of the sessions during the Peace of Amiens the effect among those who did don the required outfit was spoiled by their large variety of leg- and footwear, some sporting black silk stockings and others in worsted, some in half-boots and others in whole boots with dirty brown tops'.[12] They were obviously sheepish, too, about the ceremonial imposed on them. They were supposed to assemble in an ante-chamber and enter the hall in procession, heralded by military fanfares and accompanied by the sound of drums beating to arms; but in fact half of them sloped in beforehand, and a visitor from Germany said that even the music was out of tune.[13]

At a more serious level, it is difficult to determine what they thought about Napoleon's persistent search for glory in war. Discussion among the legislators was confined to such peripheral matters as votes of thanks destined for Napoleon, and celebrations in which they were called upon to take part. For the early years there are full reports of the proceedings of the Tribunate, but the committees which took over in later years held their deliberations behind closed doors and only the final speeches were published. For much of the time it is necessary to resort to indirect evidence, such as the terms in which Napoleon's spokesmen from the Council of State saw fit to address the assembly when presenting bills. Napoleon himself made a carefully laundered speech, once a year from 1804 onwards, when having become Emperor he opened parliament in person in imitation of the King of England.

Though the evidence is scanty, however, it seems to point unmistakably to the conclusion that the vast majority of members of parliament did not appreciate Napoleon's ambitions. They were anxious to safeguard France from the foreign powers which had threatened to restore the *ancien régime*, and for this purpose they believed it necessary to secure the natural frontiers for which France had fought since 1792; but they were no longer animated by the crusading zeal which had inspired revolutionary governments to establish sister-republics beyond the frontiers. One of the earliest assurances that Napoleon's government felt obliged to give parliament was that all desire to interfere in the affairs of the continent had been abandoned when it took office at *brumaire*.[14] There is no sign that the deputies shared Napoleon's enthusiasm for the virtues such as loyalty and self-sacrifice that were engendered by war, and they certainly had no desire to see his *régime* in France, which they supported and admired, undermined by costly and hazardous adventures.

When Napoleon came to power at the end of 1799, France had already beaten off the invading armies of the Second Coalition, and there was every chance that the Habsburg Emperor would make peace on the basis of the natural frontiers if France restored to him the Cisalpine republic she had formed in Lombardy. As Napoleon had no intention of doing so, his task in 1800 was to persuade parliament that the independence of a sister-republic was worth fighting for. He tackled it by presenting England as the hidden enemy. England had persuaded the Emperor to make these unacceptable demands. It was England that wanted the downfall of the Cisalpine republic, because she wished to see re-enacted there the ghastly reprisals that had followed the downfall of the Neapolitan republic a year earlier – an episode in which her navy had played a despicable role. Her next victim would be France herself, for the English ministry had declared that it would not end the war until the *ancien régime* was restored.[15]

Such a presentation was calculated to revive the horror which Frenchmen had always felt for British methods of warfare during the revolutionary period, when the British government had sent agents and subsequently landing forces to France to aid royalist rebellion – a procedure which aroused much the same distaste as modern terrorist activity. The First Consul, parliament was reminded, had put an end to foreign-inspired rebellion at home; one final effort was called for abroad. 'The trumpet of war sounds for the last time: it calls not for carnage but for peace!'[16]

Although Napoleon had put out a call to the troops to rally behind his leadership in the forthcoming Italian campaign, he pretended for parliament's sake that he had no intention of taking an active role.[17] The deputies were not misled, however. A few malcontents were pleased to think that he might never come back: they took to meeting regularly with Sieyès in a café at Arcueil to discuss the possibility of a *coup d'état*. A bigger group, however – possibly as many as fifty – were genuinely worried as to what might happen to France if her new-found saviour were to be killed; at the risk of being confused by the police with the malcontents, they also met regularly, in a different café, to discuss contingency plans.[18] At the news of Napoleon's victory at Marengo the Tribunate expressed relief and delight,[19] but its members were even more jubilant when Moreau's outstanding success at Hohenlinden resulted in an armistice. 'Peace! O celestial peace! Fix for ever your sojourn among us!' cried an enraptured deputy in the *Corps Législatif*.[20] The subsequent treaties of Lunéville and Amiens were so assured of a welcome that Napoleon chose the very day on which the latter was submitted to the legislature to inaugurate in the Tribunate the proceedings that were to make him First Consul for life.[21]

What, then, was parliament likely to think when war between Britain and France was renewed only a year later? Curiously enough, the news seems to have been accepted without too much anxiety, perhaps because Napoleon gave the impression that the conflict would be a limited affair. England this time was accused not of trying to destroy France's institutions but of trying to destroy her economic potential by curbing her overseas expansion.[22] England wanted to dominate world trade; England thought that only she had the right to build up empires in India, and so on. There had indeed been much anxiety in Britain over France's economic ambitions, and Napoleon was able with complete confidence to lay before parliament the correspondence that had taken place between Talleyrand and the British ambassador, for it could reasonably be argued that once war appeared to be inevitable Britain had been more intransigent in her attitudes than France. Even so, the President of the *Corps Législatif* advised patience and generosity: France must not be the one to declare war.[23] Anxiety would doubtless have been greater if fighting on the continent had been at issue, but Napoleon assured the deputies – and went on assuring them for the next eighteen months – that he had no intention of extending the conflict. 'I do not wish to increase the territory of France', he declared at the opening of the session in December 1804, 'I have no ambition to exercise greater influence in Europe.'

It was two and a half years after the renewal of war with Britain that Councillor Mounier reported to the Tribunate that Austria and Russia were moving troops against France. Interestingly, these two powers were no longer presented as despots against whom France had an ongoing crusade but as natural allies of France in the fight for the economic independence of the continent – seduced by British gold into deserting that alliance.[24] In 1804 Britain had declared a blockade of all France's channel ports, and from then on Napoleon never ceased to present himself as the champion of Europe's economic identity and of freedom of the seas.

The great victories of Ulm and Austerlitz were celebrated by Napoleon along with parliament's completion of the Civil Code, and the Peace of Pressburg which followed them was described to parliament as France's final reckoning with the continent. The Emperor, Councillor Champagny declared in the *Corps Législatif*, had 'épuisé la gloire militaire'; there would be no more conquests.[25] Parliament was not meeting when in October of that same year French troops were launched against Prussia and went on to fight a bitter winter campaign against the Russians in Poland. The only indication of the deputies' sentiments that has survived was given when a group of them was received by the Empress Josephine, and the smooth-tongued President of the *Corps Législatif*, Fontanes, commiserating with her on the absence of the Emperor, began his speech with the enigmatic words, 'The bravest of peoples is sometimes tempted to think it has too much glory'.[26] The ambitious Treaty of Tilsit produced none of the euphoric outbursts in parliament that had greeted the Peace of Amiens; and Councillor Cretet, in his report on the state of the nation, promised the legislature that from now on Napoleon would devote himself to domestic affairs.[27]

The promise was not carried out, for Napoleon very soon marched into Spain to establish his brother Joseph on the throne. When, shortly afterwards, he wrote back suggesting that the deputies might raise a triumphal arch in his honour on the heights of Montmartre, the President took the precaution of clearing the public galleries before the proposal was discussed in the *Corps Législatif*. The required resolution was passed, but no arch ever materialised.[29]

The question now arises, if it is correct that the deputies disliked Napoleon's war-mongering, why did they wait so long before trying to curb it? For it was not until December 1813, when Napoleon had been defeated in Russia and defeated in Germany, and the allies were poised for the invasion of France, that a commission of the *Corps Législatif* told him plainly that he should issue a statement signifying his readiness to accept peace on any terms short of the sacrifice of French territory. In November the allies had unofficially offered peace on the basis of the natural frontiers, and Napoleon had procrastinated. On 4 December they had come up with a vaguely worded Declaration at Frankfurt, stating that they would guarantee to the French Empire 'an extent of territory unknown to France under her kings', and Napoleon had so far not replied. The legislature now insisted that he should accept this offer, and if the allies proved to be insincere, and France had to fight for the defence of her frontiers, he must rouse the nation to a great patriotic effort by promising to restore to the French people their full rights and liberties.[30] Napoleon took the whole report not only as an insult to himself but as a threat to morale at a time of emergency; and he promptly dissolved parliament, saying that it was dangerous to leave such men in session whilst he went to fight the enemy.[31]

The members of the *Corps Législatif* were afterwards proud of their stand, but they have been criticised harshly by historians, both for not taking it earlier and for taking it now when Napoleon had his back to the wall. They have been accused either of cowardice in the face of the almighty warrior, and of screwing up their courage to criticise him only when they were terrified for their own skins if the enemy conquered, or of greedily taking all they could get from Napoleon whilst he was successful and turning against him when he was in difficulties. There is doubtless something to be said for these condemnations. It is hardly likely that the deputies had at any time been actually frightened of Napoleon, for the worst that had happened, even to the Tribunes who criticised his laws severely in the early years of the *régime*, was that they had been gradually manoeuvred out of their seats in parliament and offered jobs in the civil service; but there had always been some who had fallen over themselves to flatter the great man, in the mistaken belief that this was the road to high office. There were probably many others who had procured seats in parliament simply because of the money (they drew a handsome salary for very little work) and who saw no reason to exert themselves until things got desperate. It is only fair, however, to recognise that there were problems which made their behaviour less reprehensible.

Not least of these was the difficulty anybody would have encountered if trying to organise concerted opposition. The composition of the assemblies changed considerably from year to year, since membership was renewed annually by fifths, and the electoral system was so complicated that it was unlikely that any of the outgoing members would ever be returned again. Moreover from 1809 onwards, as more and more territory was annexed to the Empire, more and more foreign deputies arrived to take their seats: by 1811 a third of the members were foreign born. There was no party organisation through which newcomers could get to know each other: such was the general disapproval of anything approaching 'party' or 'faction' that members were not even allowed to choose where they would sit on the benches – they drew tickets for places like students entering an examination hall. The length of time they were together in Paris grew shorter and shorter as there was less and less fundamental legislation to be dealt with: in 1800 the session lasted four and a half months, but by 1813 it lasted a mere five and a half weeks. There was no fixed date or even customary date which could be anticipated for the opening of the session: it was known to commence in August, February, October, December and June. Napoleon was always loath to fix a date until the last minute, and even then he often changed it. Not until 1813 did circumstances work to some extent in the deputies' favour. From 1804 Napoleon had always insisted on opening parliament himself, which meant that the 1812 session had had to be delayed until he got back from Moscow. After some procrastination it finally began on 14 February 1813. It lasted only thirty nine days, but by a fluke it was the same deputies (minus a good many of the German members, whose constituencies were in the process of breaking away from the Empire) that met again at the end of the same year for the session of 1813. For neither occasion had Napoleon found time to order new elections, so the members who should have retired at the end of 1811 and 1812 were allowed to continue their service. Moreover, on this final occasion, Napoleon's letter deferring the opening date arrived too late to prevent the deputies from leaving home; they consequently found themselves kicking their heels in Paris for more than a fortnight, and the more determined politicians among them were able to initiate plans.

This sort of muddle and improvisation were characteristic of the Napoleonic *régime*. Of more lasting interest is the difficulty that the deputies were placed in by a constitution which had been based on the principle of the separation of the executive and legislative branches of government. Separation of powers had been one of the most cherished principles of the Revolution. In 1799 a deviation was made from it in order to give Napoleon the initiative in legislation (such was his reputation as a legislator, even at this early stage, that his hand was believed to be necessary in order to get much-needed laws that the revolutionaries had been discussing for years on to the statute book); but in other respects the principle was maintained. Ministers were not chosen from among the members of parliament; they did not sit in parliament; and they could not be made to resign by parliament. The system was believed to have mutual advantages, for whilst ministers could not be controlled by the legislature, the legislature could not be controlled by the executive power. It would have been just as difficult for Napoleon to build up a party of supporters in the *Corps Législatif* as it was for the deputies to organise an opposition.

At first sight it would appear that the legislature was going to have no power at all to criticise the government, but as usual in such situations there were loopholes and ambiguities, and much depended on what the deputies made of them. One of the loopholes in the early years was that treaties of peace and alliance had to be presented as bills and endorsed by the legislature. Napoleon had disliked this requirement from the moment of coming to power: how could a government, he complained, responsible as it was for negotiating with foreign diplomats, do so effectively unless it was known to have the last word? Nevertheless, at the end of 1801, he dutifully laid before the *Corps Législatif* the various treaties that together were to constitute the Peace of Amiens. Not one of them got a unanimous vote. The treaty with Naples was denounced in the Tribunate as condoning the Neapolitan king's wartime atrocities, and the treaty with Russia raised a furore because it inadvertently referred to French citizens as 'subjects'.[12] It was obvious, of course, that the bills were going to pass, which they did with huge majorities; but one or two dissidents were able to make their gestures. Napoleon was furious, and in the constitutional amendments which followed his promotion to the life consulship he altered the procedure so that in future he merely had to 'communicate' treaties to the Senate. As a result the legislature had no formal chance to discuss the Treaty of Tilsit, which in 1807 so enormously extended France's power over Europe.

Even Napoleon, however, would not have dared to take away from the legislature the right to endorse declarations of war, nor in his better moments would he have thought it proper to do so. In 1803 he was preparing to declare war on Britain, and it was for this reason that he laid before the legislature the letters that had passed between Talleyrand and the British ambassador. In the event, Britain stole a march on him and declared war on France. Napoleon was annoyed at losing the advantage as to timing, but appreciated the propagandist advantages, and for many years thereafter he took care always to present the enemy as the aggressor. In 1805 it was the Austrians that had taken the initiative by moving troops into Bavaria: 'They are attacking our ally; they have begun the war!'[33] In 1808 it was the British who had sent troops to Spain to help the rebels before French troops marched in: 'Stress the vexatory actions of England,' Napoleon wrote in the notes he prepared for his speech to the *Corps Législatif*, 'stress that we are only seeking reprisals.'[34] In 1814 the Senate declared that Napoleon had forfeited the throne by 'beginning a series of wars without the consent of the legislature', but if the case had gone to court it would probably have been discovered that the only occasion on which he could be faulted was when his troops crossed the Niemen in 1812 without parliament's permission for a declaration of war on Russia.[35]

It would have been possible at any time for the legislature to question how far Napoleon was justified in committing French troops abroad for purposes of defence. Article 47 of the constitution assigned to the government the duty of providing for the defence of the country from foreign attack and the right to dispose of the forces of land and sea, but did these powers, wide though they were, allow Napoleon to send forces off to Germany to protect the territory of a protégé, or to Spain to establish the rights of a puppet king? Napoleon's legislatures have not been alone, however, in their reluctance to explore their rights in the matter of defence. Even in the USA the Supreme Court has been said to approach this topic 'with all the care attributed to porcupines making love' and Congress to have challenged the government only when the electorate showed concern.[36] In Napoleonic France, members of parliament had reason to believe that the constitution of 1799, with all the wide-ranging powers it gave to the government, was popular in the country. Had it not been endorsed by an enormous majority in a plebiscite? Moreover the constitution of 1799, unlike any such document before or since, actually named the person who was to wield those powers. Napoleon Bonaparte could with some justification claim, as he frequently did, that he represented the people more directly than the politicians, who were not a popular breed – as became patently clear when in 1807 he abolished the Tribunate without a murmur from the public.[37]

The tide of popular favour did not begin to turn until the Spanish War had been in progress for some years. Up to that time, even conscription could be regarded as not much of a burden.[38] The basis of the call-up was still the revolutionary Loi Jourdan, which allowed plenty of room for manoeuvre; as long as there were more men available than Napoleon required, there was the opportunity to ballot for places, and for men with money to buy substitutes. Not until 1812 was the leeway used up. In 1813, rules would go by the board – previous classes would be combed out, future classes anticipated, and men of less than regulation height sent to the front. The time for the legislature to tackle Napoleon on his use of his authority would thus have been at the end of 1812, when he returned defeated from Russia bent upon raising another huge army to fight in Germany. Unfortunately the *Corps Législatif* had by then long since lost control of the recruitment process. In September 1805 Napoleon had needed troops quickly for the campaign in central Europe, and as parliament was not sitting he had applied to the Senate.[39] The latter, whose role was to act as a watchdog of the constitution, had decided that it was competent to sanction demands that lay well within the limits of the law; and from then on Napoleon never applied to the legislature again.

There remained finance. Napoleon even before he came to power had said that finance should be entirely in the control of the government, for how could it govern without money?[40] He nevertheless accepted a clause in the constitution of 1799 which obliged him to 'direct the income and expenditure of the state in accordance with an annual law determining the amounts thereof'. For some years he tried to escape from its stranglehold by submitting his requests piecemeal and claiming that they added up to one financial law, a manoeuvre which impressed many of the deputies with his reluctance to spend money; but fortunately a group of clear-sighted Tribunes foresaw the danger and stood out successfully for the letter of the law.[41] Had they not done so, parliament would have lost control of the financial weapon exactly as it lost control of recruitment. Ironically, however, this same group of Tribunes, acting in good faith, enabled Napoleon to make a partial escape from parliament by another route. The main source of the government's income at the outset of the *régime* was the land tax, all indirect taxes having been abolished during the Revolution. For economic reasons Napoleon was now encouraged to lighten the burden of the land tax by re-introducing consumer taxes on so-called luxury items, mainly drinks and tobacco.[42] This provided him with a source of income easily controlled by civil servants. At the same time, as soon as war broke out, parliament agreed that control of customs and excise was a police concern, intimately connected with France's security against foreign enemies, and Napoleon must be allowed to raise them or lower them as necessary.[43] These two provisions, combined with the fact that enormous sums of money were exacted from conquered countries, meant that Napoleon was able for many years to announce that he required no extra taxation from his subjects.[44]

Again the crucial moment came after the Moscow campaign, when outside resources were no longer available and new expedients had to be found. The deputies who arrived in Paris for the session which began on 14 February 1813 seem to have been aware that a decisive point had been reached: yet they failed to rise to the opportunity. After all, parliament had not met for eighteen months, and nobody had known whether it would ever meet again. Something must be attributed, too, to the charismatic effect which Napoleon's personality was capable of producing upon anybody who encountered it in a crisis. The speech he made at the opening of the session was criticised by people who read it afterwards in cold blood, but those who heard it at the time were electrified.[45]

The financial bill which formed the only business of the session was a surprise in that Napoleon once again announced that he did not need to increase the direct taxes. He proposed to make up the shortfall in his finances by confiscating the property owned by municipalities throughout France in return for government bonds.[46] A committee of the *Corps Législatif* presided over by a businessman from Rouen accepted the proposal as economically sound, and the bill passed by 303 votes to 26.[47] Only at the end of 1813, after another disastrous campaign, could Napoleon see no alternative to raising the land tax. He tried to wriggle out of putting this to the legislature by arguing that approval was needed only for new forms of taxation, not for higher levels of existing taxes, but Cambacérès, Napoleon's deputy in home affairs and the man most closely in touch with opinion in the legislature, persuaded him that this would be fatal to public morale in the coming campaign.[48] Napoleon had always believed that it was his especial duty to preserve the landed interest and that only a dire emergency could justify him for imposing burdens upon the landed classes. It was in an effort to persuade the *Corps Législatif* that such an emergency had now arisen that he took the step, unprecedented since 1803, of laying before a committee of its members the documents relating to his negotiations with the allies since the Battle of Leipzig. He claimed that the documents proved he had done everything in his power to obtain an honourable peace, but the deputies did not agree with him. He responded to their implied criticisms by immediately dismissing the legislature and mounting the campaign for the defence of France on a mixture of illegal taxes, promissory notes, and funds from his private treasure.

There has been no intention in this paper of exonerating Napoleon's parliaments completely from the criticisms so often levelled at them. They must take their share of blame, along with the French people, for doing less than they might have done to put a stop to unnecessary bloodshed. Before condemning them too harshly, however, it may be useful to remember the lesson that Jane Austen's heroine had to learn when at Northanger Abbey she left the realm of fiction for that of real life: namely that circumstances can to some degree alter cases, and that the conduct of members of a civilised nation is generally neither good nor bad but a mixture of both.

NOTES

1   G. Lefèbvre, *Napoléon* (Paris 1965).

2   Anita Brookner, *Jacques-Louis David* (London 1930), pp. 168–9.

3   For a detailed account, see Irene Collins, *Napoleon and his Parliaments, 1800–1815* (London 1979).

4   Napoléon Ier, *Correspondance de Napoléon Ier, suivie des oeuvres de Napoléon à Sainte-Hélène* (32 vols, Paris 1858–70), no. 8324.

5   Ibid., nos. 9467, 14463; *Archives parlementaires: recueil complet des débats législatifs et politiques des Chambres Françaises de 1800 à 1860* (Paris 1962–7), 1 Jan., 11 May 1806; 19 Nov. 1809; 22 Jan., 1809; *Journal du Soir*, 15 May 1806.

6   P.P. De Ségur, *Histoire et Mémoires* (7 vols, Paris 1873), iii, pp. 296–8.

7   Napoléon, *Corr.*, no. 10412.

8   Napoléon Ier, *Lettres inédites*, L. de Brotonne (ed.) (2 vols, Paris 1903), ii, 688.

9   M. Guerrini, *Napoleon and Paris: Thirty Years of History* (Paris 1970), pp. 235, 306.

10   Collins, *Napoleon and his Parliaments*, p. 141.

11   F.V.A. Aulard, *Paris sous le Consulat: recueil de documents pour l'histoire de l'esprit public à Paris* (3 vols, Paris 1912–23), 12, 17 Dec. 1801; 9, 14 Feb., 1 Mar., 14 Apr., 12 May 1802.

12   H.R. Yorke, *France in 1802*, J.A. Sykes (ed.) (London 1906), p. 207.

13   A. Laquiante, *Un hiver de Paris sous le Consulat, 1802–3* (Paris 1896), pp. 369–70.

14   *Arch. parl.*, 18 ventôse an VIII.

15   Ibid., 18, 28 ventôse an VIII.

16   Ibid., 17 ventôse an VIII.

17   Napoleon later said that this prevarication was necessary because the constitution forbade the First Consul to lead an army in the field; but there was no such clause in the constitution.

18   *Archives Nationales*, AF IV 1329.

19   *Arch. parl.*, 2 messidor an VIII.

20   Ibid., 13 nivôse, 24 pluviôse an IX.

21   Ibid., 16 floréal an X.

22   Ibid., 20 May 1803.

23   Ibid., 14 May 1803.

24   Ibid., 24 Sept. 1805.

25   Ibid., 5 Mar. 1806.

26   Ibid., 5 Feb. 1807.

27   Ibid., 24 Aug. 1807.

29   Napoléon, *Lettres inédites*, i, p. 250; *idem, Corr.*, no. 14510; J.J. Cambacérès, *Lettres inédites à Napoléon, 1802–14*, J. Tulard (ed.) (2 vols, Paris 1973), 23, 26 Nov. 1808.

30   *Arch. parl.*, 19, 21, 22, 23–7 Dec. 1813; J.H. Lainé, *Rapport de la Commission extraordinaire fait au Corps Législatif le 28 décembre 1813* (Paris 1814).

31   A.J.M.R. Savary, *Memoirs of the Duke of Rovigo, Written by Himself* (4 vols, London 1828), iii, pp. 179–80.

32   *Arch. parl.*, 26–30 Nov., 6–10 Dec. 1801.

33   Ibid., 24 Sept. 1805.

34   Napoléon, *Corr.*, no. 14394.

35   *Arch. parl.*, 3 Apr. 1814.

36   B.M. Blechman, *The Politics in National Security: Congress and U.S. Defense Policy* (Oxford 1990), reviewed by Geoffrey Marshall in the *Times Literary Supplement*, 1 Mar. 1991, p. 6.

37   *Arch. parl.*, 19 Nov. 1808; *Moniteur*, 15 Dec. 1808; Napoléon, *Corr.*, no. 15978; C.F. Méneval, *Mémoires pour servir à l'histoire de Napoléon Ier* (3 vols, Paris 1894), ii, pp. 123–4.

38   Alan Forrest, *Conscripts and Deserters: the Army and French Society During the Revolution and Empire* (Oxford 1989), pp. 41–3.

39   *Arch. parl.*, 24 Sept. 1805.

40   J.E. Howard, *Letters and Documents of Napoleon* (London 1961), p. 202. Napoleon also told the Council of State on 8 Oct. 1808, 'It must not be within the power of a *Corps Législatif* to bring government to a halt by refusing supplies.' See J. Pelet de la Lozère, *Opinions de Napoléon* (Paris 1833), p. 147.

41   *Arch. parl.*, 13, 19, 22–3 ventôse an VIII; 17 nivôse an IX; 14 Feb. 1804.

42   Ibid., 14 *floréal an* X; 14, 23 Feb. 1804; 14 Apr. 1806.

43   Ibid., 24, 27, 29 *floréal an* X.

44   Ibid., e.g. 25 Oct. 1808; 3 Dec. 1809; 15 Jan. 1810; 16 June 1811.

45   H. de Noailles, *Le Comte Molé, 1781–1855* (6 vols, Paris 1922–30), i, pp. 165–8; V. de Chastenay-Lanty, *Mémoires de Mme de Chastenay, 1771–1815* (2 vols, Paris 1896), ii, pp. 223.

46   *Arch. parl.*, 14 Feb., 11 Mar. 1813.

47   Ibid., 20 Mar. 1813.

48   Napoléon, *Corr.*, no. 20833; Cambacérès, *Lettres inédites*, 4, 9 Nov. 1813.

EQUALITY AND THE THREAT OF WAR
IN SCANDINAVIA, 1884–1905

Ida Blom

The question of equality

War – or the threat of war – may be used as a means to influence decisions in political matters having no direct implication for the question of war or peace. Giving – or withholding – support for war policies, be they offensive or defensive, may be a useful way of bargaining about other political issues. This was by and large the effect of the threat of war in Scandinavia during the twenty-one years between 1884 and 1905.

Among these other political issues, this paper shall concentrate on three issues, all closely linked with the question of equality, and all three a matter of contention in most of the Western world. *Democratisation* built on the assumption that human beings had a right to equal access to political power, irrespective of class. *Feminism* may be defined in a number of ways. It is here seen as a policy to obtain equality between women and men, with a special view to the vote. For dependent nations, *nationalism* was an expression of the hope of being recognised as equal to other nations, obtaining independent national political institutions.

In all three respects, 1884 was an important year in Norwegian history. This year saw the first organisational expression of the democratisation process. Two political parties, the Liberal Party and the Conservative Party were created, and male suffrage enlarged from a narrow base laid in 1814. Three years later the Labour Party was born. Women also started organised political activity in 1884, by forming the Norwegian Association for the Rights of Women (*Norsk Kvinnesagsforening*). This organisation was an instrument in the fight for wider access to education and paid work, as well as for a more independent status for married women. Although open to male and female members alike, the association was seen as a female parallel to the Liberal Party. The claim for female suffrage on the same conditions as men enjoyed led in 1885 to the formation of the Association for Women's Suffrage

(*Kvinnestemme-rettsforeningen*) partly overlapping in membership and leadership with the Association for the Rights of Women. This new association admitted no men. The two associations very closely mirrored the division between a moderate and a radical fraction within the Liberal Party.[1]

There were no similar Conservative feminine organisations, but, as we shall see, Conservative women were politically active within organisations closely related to the Conservative Party.[2] As far as the Labour Party was concerned, in 1894–5 women's associations were formed in the two main towns, Kristiania (Oslo) and Bergen, in support of the local party organisation. In 1901 the Norwegian Labour Party's Women's Organisation (*Arbeiderpartiets Kvinneforbund*) was created, joining the two women's associations and a number of female trade unions under one hat as a national organisation for working class women.[3]

Around the turn of the century, men and women of the bourgeoisie as well as of the Labour class had built up political organisations to serve in the fight for – or against – democracy and feminism, understood as political formal equality between classes and between women and men. But 1884 was important in another respect as well. The year marked the end of a protracted fight between supporters and opponents of the royal veto and of the principle of parliamentarism.[4] The formation of a Liberal government in June 1884 sealed the defeat of the Conservatives, who had staunchly supported the King in his fight to retain the royal veto as well as in his right to choose his government without considering the political balance in parliament. The fight had been long and hard, even to the point of the King mobilising the army against the opposition and exploring the possibility of assistance from the German Kaiser in case of an armed conflict. On the part of the opposition, popular voluntary rifle clubs had sprung up all over the country, to all practical purposes an army under the command of the Liberals. The threat of war never became more than a threat, but the situation marked the popular attitude to the army for the coming decade. The army, and especially the officer corps, were seen as the King's men, not the men of the democratic opposition. Appropriations for defence, army as well as navy, were minimised whenever possible.

This conflict to a large extent was also a conflict between nations, the Norwegians and the Swedes. Since 1814 Norway had been in a political union with Sweden, much like the one organised in Austria-Hungary in 1867.[5] Two parliaments, two capitals, two armies and navies, but one King, residing mostly in the Swedish capital, Stockholm, and responsible for foreign policies as well as for defence. It was easy to perceive the King as Swedish, more than

Norwegian. The foreign minister as well as most of the diplomatic corps consisted of Swedes, although nothing formally prevented Norwegians from obtaining these posts. However, in 1885, the result of a change in the Swedish procedures for decisions in foreign policy matters, implicitly made the Norwegian government accept that the foreign minister *would have to be* a Swede. This created an uproar in the Norwegian parliament, and resulted in a claim for two foreign ministers, a Swede and a Norwegian, as well as for separate national representation abroad. A political conflict over national equality had been born, that would end with the abrogation of the Norwegian-Swedish union in 1905.

In 1884–5, then, foundations for democratisation, for feminism and for nationalism were laid. The coming twenty years would see further growth within all of these three areas. The question of equality was at the heart of the matter – equality between classes, between men and women, and between nations. A complicated process of cooperation and conflict between proponents of the various ideologies led to full male suffrage in 1898 (part of formal political equality between classes), and to the abrogation of the political union with Sweden in 1905 (equality between nations). The first steps were also taken to create formal political equality between men and women. Limited female suffrage for local elections was accepted in 1901, and for national elections in 1907. In 1913 women obtained general national suffrage.

At crucial moments the threat of war, and hence attitudes to the question of strengthening defence, played an important part in this development. This paper will analyse three different lines in national policies as responses to the threat of war, three lines followed by Conservative, Liberal and Socialist women respectively, and with the secondary effect of advancing feminist claims for broadening democratisation to encompass women's suffrage.

The 1895 crisis and the rebirth of a strong defence

The Liberal Party – from 1887 with the support of the Labour Party – was strongly against strengthening defence in any way and tried to solve conflicts within the union through negotiations. But although the dispute about the nationality of the minister for foreign affairs and of the diplomatic corps was replaced by a claim for a Norwegian national consular service – a conflict at a lower level – negotiations dragged. A Liberal initiative to solve the problem simply by considering decisions about the consular service as part of Norwegian *domestic politics* and declining further Norwegian-Swedish negotiations, failed in 1895. The Swedish parliament reacted strongly to this proposal by augmenting war-appropriations. The Secret Council' (*sekreta utskottet*) – which was only to meet if a war was threatening – was summoned. The foreign minister, friendly to the Norwegians, was replaced by a more belligerent person and the idea of coercing the Norwegian parliament to accept a closer union of the two countries was aired. Rumours current at the time suggest that support was again looked for in Germany. In any case, Swedish preparations for a military conflict made the Norwegian parliament climb down and accept continued negotiations.[6] This crisis made the Liberal Party, which had led the nationally assertive policy, change its view on defence matters. Appropriations to strengthen the Norwegian forces were no longer obstructed.[7] In the question of defence the Liberals now joined hands with the Conservatives.

Conservative politicians and military men friendly to the union with Sweden had already in 1886 established 'The Norwegian Defence Association' (*Norge Forsvarsforening*), possibly as a counterbalance to the popular Rifle Clubs.[8] This association saw it as its main aim to strengthen the popular will to spend more money on defence, by spreading information and petitioning parliament and government.

In the autumn of 1889, one of the Kristianian papers brought the news that 'several of the ladies of our town' had taken an initiative to pursuade the Defence Association to start fund raising for defence matters. But although substantial amounts had lately been donated for the association to administer, the association did not see fund raising as part of its work. However, there was nothing to prevent the leaders of the Defence Association from assisting the ladies in organising their activity, although keeping a certain distance by stressing that 'the honour to have had this thought, belongs to the ladies, and so does the honour of the result.'[9] On 16 November a Women's Circle was created within the Kristiania Circle of the Defence Association, to act as the central organiser of a fund raising campaign.[10]

The next day a group of 56 women published an appeal in the press:

TO THE WOMEN OF NORWAY
If the choice between peace and war in the world was dependant on our voice, the women's voice, there could be no doubt of the result. We love peace.[11]

But, according to the appeal, love of peace was the reason why women would have to prepare for defending peace. It was maintained that women had to rely on peace in all their undertakings, and that they suffered from war in a special way as they had to watch inactively the fight put up by their beloved ones. Based on these considerations, women were asked to pay whatever they could afford into a fund for a new warship. Women were also asked to create local circles to administer fund raising throughout the country.[12]

Four other women's circles were created. The Women's Krone-Collection, also in Kristiania, saw it as its task to collect smaller gifts to benefit defence and through lectures to 'counteract peace-agitation'.[13] A Women's Circle in Bergen, the second largest city in Norway, financed the strengthening of the defence of the seaward approach to the city.[14] The circles in Kristiansand and Storelvdal have also left traces in the annual reports of the Defence Association.[15] After two years the campaign had brought enough money (nkr. 600,000.00) to build 'the women's warship', a torpedo-boat named *Valkyrien*. A Valkyrie was a heathen northern war-goddess, who at the command of the main god, Odin, was said to go to the battlefield to decide who would be killed. Parliament unanimously granted a small appropriation (nkr. 55,000.00) to complete the installation of weaponry. On 17 May 1896 – the Norwegian National Day – the torpedo-boat was handed over to the navy.[16]

There was, however, no anti-Swedish element in this action.[17] What prompted it was a vague fear that some time in the future Norway might need a stronger navy, irrespective of whether defence would be a joint Norwegian/Swedish action or an independent Norwegian task.[18] The fund raising, therefore, seems to be an expression of the general conservative will to strengthen defence.

Participating as a pressure group in formulating defence policies was not a male prerogative. Although the Women's Circles may have been started on the initiative of men, the Norwegian Defence Association came to depend on the activity of women. Not only did they manage to collect the money needed for 'the Women's Warship' and the extra fortifications in Bergen, their work also resulted in a mighty boost to the organisation. The number of members doubled in two years, and by February 1891 the organisation mustered 42 circles. No wonder the central working committee sent a letter to all the Women's Circles not to end their work when the fund raising was finished, asking them to continue to spread information and recruit new members. However, with the fund raising completed, the number of members fell drastically. When resuming activity after 1905, the organisation had to be completely rebuilt.[19] The activity of the Women's Circles had been decisive in the rise of the Defence Association. And this organisation influenced the national policy of strengthening the navy: petitions to Parliament in April 1895 from 34 local circles within the organisation, were expressly mentioned in the proposal from the Parliamentary Defences Committee as a reason for appropriating extra sums to the navy.[20]

The threat of war and female suffrage

Support for the fund raising was not limited to women from Conservative quarters.[21] Women writing in the journal of the Norwegian Association for the Rights of Women also commented on the fund raising for the warship *Valkyrien*. They all wrote anonymously. What united these anonymous voices was the idea that if women were given a say in the question of war and peace, there would be no wars. 'From the day women all over the civilized world enjoyed equal political rights with men – from that day war would be banished from the earth', wrote one optimistic woman.[22]

'Sc.' – who was identified as a man[23] – saw the fund raising initiative as a good way of proving women's public spirit, demonstrating an understanding for more than 'special interests', and for the thought that women might 'contribute to the common good by other means than bearing and educating children, that she has other duties than to cook and be the "adornment of the home", ... evidence that woman finally is waking up, a promise of a richer result of our work in the future.'

'M.D.' was against the project of raising funds for a warship. As we shall see, her reasons were mainly pacifist, but she also found it wrong, seeing that most women were hardly able to survive on what they earned, or would have to ask their husbands for money to donate. But her strongest objection was that so far women had no say in decisions concerning their own welfare and – as she saw it – were ridiculed when they tried to work for the common good.[24] All the anonymous contributors in various ways connected the question of war to the question of women's suffrage, and pledged the women behind the fund raising to work for this political goal. Given that the women behind the fund raising appeal were recruited from Conservative quarters, this was a *coup*.

The Conservatives were opposed to female suffrage. If some Conservative votes could be won over and added to the traditional supporters from the Liberal party, a victory would be nearer. Feminism and national defence policies were intertwined.

In spring 1896, as the *Valkyrien* was nearing completion, the battle over extension of male suffrage at *local elections* to include all male tax-payers seemed promising for female suffrage.[25] Changes in conditions for local suffrage could be made by a simple majority decision, and did not need the two thirds majority in Parliament necessary to carry a law concerning national suffrage. The Constitutional Committee recommended that women be included in this suffrage reform. But on 29 April this proposal was defeated by 63 votes to 21. No Conservatives changed their mind, and even some Liberals defected.[26] Fund raising for defence matters had not earned women the vote and there was widespread despair among the supporters of women's suffrage.[27] Moderate members of the Association for Women's Suffrage proposed a change of tactics. The strategy of demanding female suffrage on the same conditions as for men – which in 1896 would have meant general female suffrage at local elections – should be abandoned and suffrage won step by step, first depending on the amount of tax paid. General suffrage should be recommended at a later stage. This proposal split the Association for Women's Suffrage. In 1898 the more radical members formed a new organisation, the National Association for Women's Suffrage (*Landskvinnestemmerettsforeningen*), to continue the former strategy: female suffrage on the same conditions as for men. The threat of war would be important also to this new organisation.

'Union of knowledge and means'

The next initiative to connect the threat of war with the work for women's suffrage came from women sympathising with the Liberal Party. Their actions bore some resemblance to the Women's Circles of the Norwegian Defence Association, but were still fundamentally different.

Among the anonymous voices in the journal of the Association for Women's Rights who had commented on the fund raising for the *Valkyrien*, M.D. had been negative to the whole enterprise. M.D. has been identified as Vilhelmine Ullmann, mother of one of the Liberal MPs who most devotedly fought for women's suffrage, and of one of the leading suffragists.[28] Vilhelmine Ullmann proposed other means for women to show their interest in the question of war and peace. She suggested the establishment of an organisation of women to assist casualties, in times of peace as well as in times of war.[29] She saw this as a different way for women to show their public spirit, a way rooted in women's experiences as mothers who came to the rescue of frightened children. Private means, if possible combined with public assistance, might give women the possibility of acquiring more knowledge of how to assist with casualties and nurse the wounded. The same organisation might take care of creating stores of everything needed in case of catastrophies, such as dressing-apparatus, blankets and linen.

The Association for Women's Rights was approached from various quarters to take the lead, but declined to support the idea. It was argued that an association to promote nursing of the sick and wounded in times of war, and teaching courses to the same effect already existed. This may have been an allusion to the Norwegian branch of the International Red Cross, established in 1863. But the main obstacle was that the Association for Women's Rights had no money available to secure stores of the material needed.[30]

Vilhelmine Ullman's initiative therefore had no immediate effect. But six years later an organisation with exactly the same purposes, the Norwegian Women's Sanitary Association (*Norske Kvinners Sanitetsforening*), was established. Other forces were also at work. Sanitary General J.F. Thaulow, leader of the military medical services since 1889, was also president of the Norwegian branch of the International Red Cross. It had remained a Kristiania-based organisation, limited to men of the leading social circles, and Thaulow was eager to broaden its membership. In 1893 a branch was created in Bergen, in 1894 another in Trondheim. The Trondheim branch was the first to admit women, and no sooner had a women's circle been established than the organisation started growing. A women's circle was established in Bergen in 1895. In February 1896 Thaulow gave a talk on the Red Cross to the Association for the Rights of Women in Kristiania. His idea was to establish yet another women's circle of the Red Cross.[11] Thaulow had rewarding experiences cooperating with women. He had been actively supporting the Women's Circle of the Defence Association,[12] and he had seen this organisation grow as a consequence of women's activities. He now attempted to use the same procedure for the Red Cross. Rules for the Kristianian women's circle of the Red Cross were formulated, stipulating that the circle was part of the Red Cross. The constituent meeting was called for 26 February 1896.

But the result of that meeting was not a women's circle of the Red Cross. An independent organisation, the Norwegian Women's Sanitary Association, was established. It did not limit its activities to times of war, but took up the three fields of activity already outlined by Ullmann in 1889, and also to be used for peacetime emergencies. Equally important, this organisation did not admit men. It was as closely related to the Liberal Party as the Red Cross was to the Conservative. This time Sanitary General Thaulow was not in luck when trying to mobilise women for his own purposes.

The women involved were Frederikke Marie Qvam, Randi Blehr, Cecilie Thoresen Krog, Margrethe Vullum and Pylle Horst, all connected by family ties to prominent Liberals and involved in feminist organisations.[13] The same was true of most of the women soon to be active in establishing branches of the new organisation in other towns.[14]

There was a clear difference in purpose between the Red Cross and the organisation planned by these women. Siding with the Liberal Party, and with its radical wing at that, the new organisation was meant as women's support to strengthen Norwegian defence in the event of a military conflict with Sweden. But the other important purpose was to assist in case of civil catastrophies, such as flooding and fires. As it was, this later became the main area of activity, and the organisation especially focused on combatting tuberculosis. The Red Cross worked for sanitary assistance in war, without taking sides in political conflicts. Its leaders were Conservative. General Thaulow presided over the Norwegian branch of the Red Cross as well as for some time over the Kristiania Circle of the Norwegian Defence Association.[35] Among the board members of the Norwegian branch of the Red Cross were Admiral Koren, who at a critical stage in negotiations over union problems in 1893, was said on a personal initiative to have sent cannon and torpedo-boats to Kristiania in order to 'prevent possible left-wing disturbances'.[36]

Women trying to support the new Liberal defence policy could not cooperate with Conservative leaders of the Red Cross. Drawing on resources within the traditional role of women, the care of sick and wounded individuals, and stressing the importance for such care also in times of peace, their policy was conceived to fit in with another new line in the politics of the Liberal Party, an active social policy. In 1891 the Liberals had hoisted the flag of social policies aiming at winning voters who flocked to the newly established Labour Party. Finally, the founders of the Norwegian Women's Sanitary Organisation clearly connected all this activity with the fight for female suffrage.[37] Founded in February 1896, the Women's Sanitary Organisation had no impact on the discussion on local suffrage for women on 29 April the same year. But new strategies soon emerged to build on earlier experiences.

Siamese triplets

The Women's Sanitary Organisation recruited members and created local groups at an astonishing speed. The number of members doubled within two years, and again after another seven years. By 1905 there were 51 local groups, of which 44 per cent were located in rural areas. This expansion went hand in hand with the growth of the National Association for Women's Suffrage. The Association for Women's Rights thrived in the same period and there seems to have been a very close cooperation between these organisations. In the bigger towns double or treble membership occurred. The energetic and somewhat despotic leader of the Women's Sanitary Organisation, Marie Frederikke Qvam, also headed the National Association for Women's Suffrage, and – from 1899 to 1903 – the Association for Women's Rights as well. The three organisations worked as siamese triplets.[38]

It would not have been surprising if this mobilisation had convinced all members of the Liberal Party of the importance of female suffrage. But this did not happen. In 1901 ten Liberal MPs proposed general male suffrage at local elections. A petition of female suffrage on the same conditions as for men, organised by the National Association for Women's Suffrage two years earlier and with 12,000 signatures, was ignored. Under the guidance of Frederikke Marie Qvam, however, the organisation succeeded in winning ten other Liberal MPs for an additional proposal to build on the petition. Within a very short time parliament was drowned in petitions for female suffrage on the same conditions as for men. A tantalising fight over the issue divided parliament. Repeated votes split the Liberals as well as the Conservatives. Probably to counteract the effects of general male suffrage, some of the Conservative MPs supported female suffrage for the first time, although limited by census provisions. The final vote in the *Lagting* carried this proposal 15 votes to 14. Female local suffrage on a very limited franchise (based on a taxable income of nkr400.00 in towns and nkr300.00 in rural areas) was a fact. The event was greeted as a remarkable victory, even if full female suffrage was defeated.[39]

The narrow victory was due mainly to the extremely active intervention of the three women's organisations, at the time united under the leadership of one women, Frederikke Marie Qvam. But the next step, parliamentary female suffrage, had to wait until after the abrogation of the union with Sweden. This time, also, a threat of war was involved and nationalism mobilised simultaneously women of all political parties. Before we discuss this final crisis, the third way chosen by women when confronted with the threat of war will be explored.

Women against war

The fund raising for the women's warship was not greeted with acclamation by all Norwegian women. As had been observed, as soon as the appeal to collect funds was published, the Liberal Vilhelmine Ullmann raised her critical voice.[40] To her, building warships was not the way to solve national conflicts. She recommended armies to go on strike, to refuse to kill or be killed. Such a strategy, she thought, might lead to a progressive paradise, where cooperation, intelligence, inquisitive science, enterprise, hard work, idealism, tolerance, high sexual morals and justice would reign, in short to everything 'the women's and peace movements' stood for. Even if she admitted this to be a fantasy, she found that at least such a fantasy would create joy, hope and fruitful thoughts. At the very least, it was in direct opposition to the idea of a 'women's warship'. For Vilhelmine Ullmann, support of 'the great peace movement that is touching all European countries' was a citizen's duty and the safest protection of national frontiers.[41]

Clearly, women – just like men – were divided on the question of war and peace. But Ullmann – like those Liberals who had accepted the fund raising – also talked of an extra reason for distancing women from defence policies:

they had no suffrage rights. Ullmann would qualify as what has been called a feminist pacifist',[42] and she was not the only one. 'All this male uproar ... serves no purpose ... and leads only to ruin and shame', maintained the female speaker at one of the yearly Labour Party demonstrations for women's suffrage on 17 May 1898.[43] Birgitte Weltzin Sørensen argued that the conflict with Sweden should be solved by strengthening Norwegian democracy, not by stepping up armaments. Female suffrage, she said, would be:

the strongest armament we can find in order to stand up to our neighbours ... Let us therefore curtail a little military armaments ... and let us arm ourselves with female suffrage so that our nation can stand secure, confident of the responsibility felt by all its inhabitants and the invincibility of our homeland.

Peaceful means, and among them the extension of democracy through female suffrage, was for her – and for a greater part of the Labour Party – the way to solve the national conflict. The following year, at the same demonstration, she talked along the same lines. Women's political objectives were peace, cooperation and justice, unlike what she qualified as men's objectives: war and blood. The Swedes were friends and fellow workers, and Norwegian female suffrage could be a means to assist the Swedes in their fight for democracy and social justice. As late as 17 May 1905, in the middle of the acute national crises, the speaker at the Labour women's demonstration for female suffrage, Anna Gjøstein,[44] said: 'We live in difficult times. The flames of war can easily be set alight. For this reason, women ought to have a say, if a council of war is established, so that the danger of war may more easily be averted.'

The peace-movement strategy, thus, was followed up all through the political turmoils. Some women – and some men, mostly socialists – adamantly adhered to this strategy in the face of threat of war. Stretching historical facts slightly, it may be said that this strategy won over war activism when in 1905 the Norwegian-Swedish union was finally dissolved.

'the supreme good, independence, autonomy and freedom'

After failings and pauses, negotiations about the consular services were resumed in 1903. But a proposal for parallel laws regulating Swedish and Norwegian consular services respectively still stipulated that the foreign minister would have to be Swedish. Negotiations stopped again in February 1905. The Norwegian parliament then proceeded to adopt purely Norwegian laws regulating a Norwegian consular service. The King refused to sanction these laws, and the Norwegian government resigned. On 7 June, the parliament decided that since the King had no Norwegian government, he could no longer act as King of Norway. The supreme national power was returned to Parliament and the political union was dissolved.

A plebiscite on 13 August strengthened this parliamentary decision and negotiations over how to dissolve the union were started in late August. By then the atmosphere in Sweden had become more militant. One of the conditions formulated for accepting the abrogation of the union was that the newly built fortresses along the Swedish-Norwegian frontier be dismantled. This sharply raised the political temperature in Norway. The navy was made ready for conflict and part of the army was mobilised in eastern Norway. But Social Democrats and some Liberals in Norway as well as in Sweden were against the use of force. In an internationally heated atmosphere with the Russo-Japanese war and the Morocco crisis, the great powers did not want a conflict in this peaceful corner of northern Europe. A compromise was reached. With a few exceptions, the Norwegian delegation gave up the claim to preserve the new fortifications, and the Swedish Liberal and Social Democratic proposals for a peaceful dissolution of the union carried the day. An agreement was reached on 23 September, and a new plebiscite in November decided in favour of a monarchy, against a republic.[45]

Women again participated in the turbulent political process. To bolster the decision of Parliament on 7 June, they wanted to take part in the plebiscite on 13 August. The problem was that women only had limited local suffrage. Marie Frederikke Qvam, president of both the Norwegian Women's Sanitary Organisation and the National Association for Women's Suffrage, telegraphed the president of parliament to the effect that, if women were allowed to take part in the plebiscite, 'you will get the best allies, because you will get those who deeply understand what it means to fight year in and year out against outer superior forces for the supreme good, independence, autonomy and freedom.'[46] But including women would have delayed the plebiscite and their demand was refused.

The Women's Sanitary Organisation and the National Association for Women's Suffrage now joined hands in collecting signatures for a petition to parliament, supporting the decision to dissolve the union. In a very short time almost 300,000 signatures followed the documents that were handed over to the parliamentary presidency. The newly founded Norwegian section of the International Council of Women also activitated its constituent associations, and 565 such associations — ranging from the association of women teachers to women's organisations for total abstinence and moral conduct — also sent petitions to support the government.

In this spirit of solving conflicts by peaceful means, it is not surprising that for once the gap between the women's section of the Labour Party and the bourgeois women's suffrage associations, was closed. Also Labour women expressed their disappointment at not being allowed to participate in the plebiscite and signed the petition organised by the National Association for Women's Suffrage.[47] If women were not allowed to take part in the plebiscite, their influence on nationalist politics was appealed to in the Labour Party paper *Social Demokraten*. Women were asked to 'convince sluggish husbands and feeble sons', that they had to use their vote at the plebiscite to support the dissolution of the union:

Just as you women protect your homes from within, you should demand of all men, young and old, who have the right to vote that they protect our common home, Norway, from without. All must take part in this task ... Women also have reason to demonstrate that they deserve the right to vote.[48]

The informal power women might have had over men in their family and surroundings was recognised and appealed to in a situation of crisis.

When, after the dissolution of the union with Sweden, parliament in 1906 again discussed the question of female suffrage, a host of petitions supported the claim. Some of them directly referred to what women had done in 1905: 'When last year our people faced the hardest fight for the full freedome and independence of our country, women stepped forward and showed that they lacked neither patriotism nor the spirit of self-sacrifice', said a petition from Stavanger. But it was not till the following year that the constitutional committee proposed female national suffrage along the same lines as had already been accepted for male local suffrage. Now the 1905 women's initiatives were extensively referred to, and 'the political interest and the political tact' shown by women in times of a national crisis was advanced as an important argument for finally accepting female national suffrage.[49] Parliament approved the proposal by 96 to 25 votes.

In 1907 feminism – and democracy – had won an important victory. Four years later, suffrage for women on the same conditions as for men, i.e. general suffrage, was accepted at local elections, and in 1913 the same principle was unanimously applied to national elections. Formal equality between women and men and between classes had been achieved on a limited, but important field, along with national equality between the Norwegians and Swedes.

Gender and the threat of war

It is often maintained that women, because they are women, have a special repugnance to war and a strong preference for peace. Historians of women have stressed the connection between feminism and pacifism as two ideologies both critical of the use of force.⁵⁰ However, a closer study of how organised women reacted, not just to war, but to the mere threat of war, indicates that – unfortunately – this is not the whole truth. Supporters of the theory that there is no monocausal explanation of wars, and that 'the male half of humanity' is not to be blamed indiscriminately for the many wars throughout history, seem closer to empirical findings. I would advance a few thoughts on possible connections between gender and war.

Women, as men, react to the threat of war in ways determined by class as well as by gender, and strongly influenced by what is at any given moment politically expedient. In the late nineteenth century, class distinctions might – as was obvious with the adherence of the early socialist movement to internationalism and the struggle for peace – go a long way in deciding the attitude both of men and women, broadly speaking leaving the more Conservative classes as the stronger supporters of using force as a means of solving international conflicts. This has also proved the case in this study. One of the reasons for the upper-class women's rally to fund raising for the women's warship may be found in the fact that the Defence Association was a continuation of earlier expressions of allegiance to the King and the Norwegian-Swedish union, in opposition to the popular Rifle Clubs. Similarly, the preference for peaceful means to solve conflicts, so clearly expressed by the Labour women and some of the Liberal women, may be rooted in the fear of being exposed to the control of the army as a means of maintaining national order. When, in 1895, the Liberals changed their position in defence matters and actively promoted a strong military defence, this change was loyally supported by Liberal women, whereas women and men of the Labour Party kept up their anti-military attitude.

These observations would indicate that men and women within the same class shared convictions on peace and war. But women did not choose – or were not given the possibility to choose – the same expressions as men for their attitudes. Men – as from 1898 men of all classes – could express their views through the vote. Women were denied this possibility. As a substitute, Conservative women would muster their long training in fund raising activities to express their opinion. Their support for armament was a purely economic support, in which men also participated. Liberal women created a special organisation, using traditional feminine training in care for the sick and helpless as an expression of their view, and formally, if not always in practice, explicitly barred men from taking part in their activity. But this activity was, all the same, support for the possibility of going to war. It was also very clearly connected with their fight for equality with men in political decision-making, their fight for the vote, an activity fairly constantly supported by the radical men of the Liberal Party.

A few of the Liberal women, represented in this study by Vilhelmine Ullmann, preferred flatly to refuse war as a means of solving international conflicts. This was the main line of action chosen by the socialist women, well in line with men of the same class. The socialists relied on public meetings and demonstrations to further their cause, although they shared with the Liberals the strategy of lobbying MPs for female suffrage. Men in the Labour Party, more consistently than in any other party, supported this claim, although even the socialists found male suffrage more important than female suffrage.

Finally, women as well as men used the threat of war to further other political interests. For women, the vote was at stake. The threat of war repeatedly offered them the possibility of making themselves visible in the political landscape. Their various initiatives were never made conditional on their obtaining the vote, but they extended and strengthened their organisational network and made it an effective tool in lobbying and petitioning for the vote. Last, but not least, female responses to defence policy and activity during the final national crisis in 1905, earned women a sympathetic reaction when the question of the vote was discussed in Parliament and contributed to victory in 1907.

The period 1884 to 1905, then, saw the fight and victory of three ideologies, democracy, nationalism and feminism, all striving for equality, all engaging women and men, and all intertwined. Equality between men of different classes came first, as general male suffrage was won in 1898. Next, equality between nations was reached as the Norwegian-Swedish union was dissolved in 1905. Last, but not least, a symbol of equality between women and men, equal political rights, gradually carried the day in successive parliamentary decisions between 1901 and 1913. The threat of war was of importance to this development.

NOTES

1   Aslaug Moksnes, *Likestilling eller Særstilling? Norsk Kvinnesaksforening, 1884–1913* (Oslo 1984), pp. 45–54, 72–96.

2   It was not until 1898 that Hjemmenes Vel (The Association for the Well-being of the Homes) was started in Kristiania. The organisation to some extent may be considered a Conservative women's association, although it clearly wanted to be seen as 'unpolitical' and was not involved in the fight for women's suffrage. See Kari Sletvåg, '"Vi Husmødre slutter oss sammen." – Fra Hjemmenes Vel til Norges Husmorforbund, 1898–1915' (unpublished MA thesis, Department of History, University of Bergen 1980).

Equality and the Threat of War in Scandinavia

3 Vera Espeland Ertresvaag, 'Arbeiderkvinnenes faglige og politiske organisering, 1889–1901', in Ida Blom and Gro Hagemann (eds), *Kvinner Selv … Sju Bidrag til Norsk Kvinnehistorie* (Oslo 1977), pp. 47–70.

4 For general accounts of Norwegian history in the period 1894–1905, see Bernt A. Nissen, *Nasjonal Vekst*, in *Vårt Folks Historie*, vol. vii (Oslo 1964); Jostein Nerbøvik, *Norsk Historie 1870–1905* (1st edn, Oslo 1973; 2nd rev. edn, Oslo 1986); Per Fuglum, *Norge i Støpeskjeen, 1884–1920* in *Norges Historie*, vol. xii (Oslo 1978).

5 Rolf Danielsen, 'Den norske oppfatningen av Unionen', in Grethe Wærnøe (ed.), *Unionstiden 1814–1905. Svenske-norske Myter og Realiteter: Dialog Norge-Sverige 1987–8* (Norges ambassade, Stockholm September 1987), pp. 11–16.

6 Fuglum, *Norge i Støpeskjeen*, pp. 86–93; Leiv Mjeldheim, *Folkersøla som Vart Parti: Ventre frå 1880-åra til 1905* (Bergen 1984), pp. 424–6.

7 Fuglum, *Norge i Støpeskjeen*, pp. 105–10.

8 The initiative came from younger officers and the founding meeting was held in the Kristiania Military Society. The Kristiania Circle was established at the same time. (A.J.T. Petersson (ed.), *Norges Forsvarsforening, 1886–1961* [Oslo 1961], p. 12). The board of the Norwegian Defence Association (henceforward NDA), presented at a meeting on 26 February 1891, was Colonel L. Chr. Dahl (president), cabinet member Birger Kildal (vice-president), university librarian Drolsum (leader of the working committee), Captain Keilhau and Consul I. Anderson Ars (board members), First Lieutenant H.D. Lowzow (secretary), and Staff Sergeant Kjærboe (clerk).

9 *Morgenbladet*, 12 and 14 Nov. 1889. See also Petersson, *Norges Forsvarsforening*, pp. 13–14; and Rolf Scheen, '*Valkyrien* – damernes krigsskib', in *Norges Forsvar* (1961–2), pp. 180–5.

10 The board for the Women's Circle was composed of women from the same social group as the men in the Kristiania Circle. *Norsk Biografisk Leksikon* gives information on three of these women (nos. 4, 8 and 9) and on the husbands of eight of them:

1. Emma Sophie Thaulow (*née* Heiberg), married to Major-General John Fredrik Thaulow, president of the Kristiania Circle of the NDA, 1881–97; chief of the army and navy medical services.

2. Elise Løvenskiold (*née* Wedel Jarlsberg), married to Carl Otto Løvenskiold, naval officer; prime minister in Stockholm, Apr.-June 1884; later Conservative Party MP.

3. Johanne Bolette Aschehoug (*née* Aschehoug), married to her cousin, Torkel Halvorsen Aschehoug, professor; Conservative politician, friendly to the union.

4. Elise Sofie Aubert (*née* Aars), married to Ludvig Aubert, professor; member of the government, Apr.-June 1884.

5. (possibly) Anne Marie Sofie Bang (*née* Schøyen), since 1888 widow of Lieutenant-Major Thorkild Berg Bang.

6. Bodil Christine Frøhlich (*née* Grønn), married to Theodor Christian Brun Frøhlich, Marshall of the Royal Court and personal friend of King Oscar II.

7. Hansine Amalie Halvorsen (*née* Smith), married to John Magnus Halvorsen, businessman; a Liberal who later joined a Conservative group which split off from the Liberal Party in 1907.

8. Eva Helene Nansen (*née* Sars), a well known singer, married to Fridtjof Nansen, scientist, explorer and politician; 'radical as far as the problem of the political union was concerned'; applauded the politics of 'no negotiations' in 1892.

9. Mrs Agathe Backer Grøndahl, prominent pianist, composer and music teacher.

10. Mrs Captain Hansen (not identified).

Apart from Major-General Thaulow, none of the husbands were at the same time involved in the board of the Kristiania Circle of the NDA.

11   *Morganbladet*, 17 Nov. 1889. See also 18 Oct. 1889 for an anonymous appeal to the NDA to assist women who were interested in collecting money for national defence to organise themselves; and *Nylænde* (journal of the Norwegian Association for the Rights of Women), 15 Oct. 1889, pp. 308–11.

Of the ten women active on the board of the Kristiania Women's Circle, elected on the previous day, eight had signed the appeal. Only Eva Nansen and Emma Thaulow did not. Among the other women who signed the appeal were the wives of three of the nine men on the Kristiania Circle of the NDA (Mrs Kildal, Anderssen-Aars, and Kjærbo).

12   I am indebted to Professor Rolf Danielsen, Department of History, University of Bergen, for drawing my attention to the existence of the 'women's warship'.

13   NDA, *Annual Report for 1891 and 1892*, p. 4.

14   In 1892 the Bergen Circle collected about nkr70,000 for this purpose. The defence department started work on the fortifications during the summer of 1894 (*Norges Forsvars-forenings Aarsberetning, 1891–2*, pp. 31–2; and *1893–5*, pp. 60–2).

15   *Annual Report[s] for 1893–5*, p. 75. Both these women's circles had collected money and forwarded it to the Kristiania Women's Circle.

16   *Sth. Prp.* no. 114 (1895); *Indst. S. XXVI B*, pp. 8–10; and *St. forh.* (1895), p. 2103 (parliamentary papers). See also, Scheen, '*Valkyrien*', pp. 180–5; and *Morgenbladet*, 18 May 1896.

17   At the founding of the Women's Circle the president of the Kristiania Circle of the NDA expressly stated that there was no wish to see the union weakened (*Morgenbladet*, 19 Nov. 1889). Representatives of the Women's Circle, of parliament and of the navy present at the ceremony on 17 May 1896, moreover, sent a telegram to the King to congratulate him on his successful convalescence after a serious illness (ibid., 18 May 1896).

18   Drolsum, according to *Morgenbladet*, 19 Nov. 1889, and 18 May 1896.

19   Petersson, *Norges Forsvarsforening*, pp. 13–14.

20   *Annual Report[s] for 1893–5*, pp. 19–24; *Indst. S. XXVI B* (1895), p. 9.

21   Even workers in Sagene, a working-class area of Kristiania, collected money through theatre performances and a workers' choral society. This was regarded as truly extraordinary (Scheen, '*Valkyrien*').

22   'hm' in *Nylænde*, 15 Oct. 1889, pp. 310–11. Other anonymous contributions in the journal were 'L' (15 Nov. 1889, p. 348); 'M.D.' (15 Oct. 1889, pp. 308–10, and 1 Feb. 1890, p. 39); and 'Sc' (15 Dec. 1889, pp. 375–8). 'L' did not like the idea of fund raising, preferring a special tax on all income, including women's income: 'Such a procedure has among other things the advantage of not fostering any martial enthusiasm – a very unfeminine sentiment.'

23   See 'M.D.', in *Nylænde*, 1 Feb. 1890, p. 39.

24   Ibid., 15 Oct. 1889, pp. 308–10.

25   Anna Caspari Agerholt, *Den Norske Kvinnebevegelses Historie* (2nd edn, Oslo 1973), pp. 184–90; Eva Kolstad (ed.), *Utsnitt av Lovforslag, Komité-innstillinger og Debatter i Stortinget om Stemmerett for Kvinner, 17. mai 1814–11. juni 1913* (Oslo 1963), pp. 52–7.

26   See Mjeldheim, *Folkerørsla som Vårt Parti*, pp. 332–43, for a study of the Liberal Party and female suffrage. Support was meagre in the party's local associations, and mainly to be found with a handful of leaders, married to or otherwise close to women prominent in the feminist movement.

27   Agerholt, *Den Norske Kvinnebevegelses Historie*, p. 190.

28   Her childhood name was Mina Dunker, hence the pseudonym 'M.D.'. Ullmann, separat-
      ed from her husband, was a teacher and publicist. She greatly influenced her children, the
      Liberal MP, Viggo Ullmann, a persevering supporter of women's suffrage; and Ragna
      Nielsen, co-founder of the Norwegian Association for Women's Rights in 1884, and of
      the Women's Suffrage Association in 1885, but an opponent of the group which formed
      the National Association for Women's Suffrage in 1898. *Norsk Biografisk Leksikon;* Agerholt,
      Den Norske Kvinnebevegelses Historie, pp. 73–86, 213–16.

29   *Nylænde,* 1 Jan. 1890, pp. 6–7.

30   Ragna Nielsen, in ibid., 1 Apr. 1890, pp. 97–8.

31   Synøve Bringslid, 'Norske kvinners sanitetsforening: stifting og aktivitet, 1896–1905'
      (unpublished MA thesis, Department of History, University of Bergen 1985), pp. 12–17.

32   *Morgenbladet,* 19 Nov. 1889. Thaulow, who chaired the meeting where the Women's Circle
      was formed, was married to a member of the Circle's board (see n. 10).

33   Frederikke Marie Qvam and Randi Blehr, both members of the Norwegian Association
      for the Rights of Women, of the Women's Suffrage Association, and (from 1898) of the
      National Association for Women's Suffrage, were married to the radical Liberal prime
      ministers Ole Anton Qvam and Otto Blehr respectively. Margrethe Vullum, an active
      journalist engaged in Liberal politics, was daughter of the Danish Liberal politician, Orla
      Lehmann, and married to editor Erik Vullum. Pylle Horst (*née* Pauline Elizabeth Gad),
      married to Liberal MP Hans Jakob Horst, was co-founder of the Women's Suffrage
      Association. Cecilie Thoresen Krog, the first woman to study at a Norwegian university,
      was a co-founder of both the Association for Women's Rights and the Norwegian Asso-
      ciation for Women's Suffrage, and sister-in-law of Gina Krog, the most prominent of
      the Norwegian suffragists (married to Lawyer Fr. Krog, also a co-founder of the Norwe-
      gian Association for Women's Rights). (Bringslid, 'Norske Kvinners Sanitetsforening',
      pp. 9–11).

34   In Trondheim Elizabeth W. Jahn and Mimi Spørch were connected to the Liberal Party
      both by marriage and through their Norwegian Association for Women's Rights work
      (ibid., p. 27). Conflicts between Liberals and Conservatives, as well as between the pro-
      ponents of the Norwegian Women's Health Organisation and the Red Cross, impeded
      the founding of a Bergen branch. This was delayed until 1898 when conflicts between the
      two parties were less conspicuous and the idea of a new moderate alliance to combat the
      socialists was forming in local politics (ibid., pp. 35–46).
           In Kristiansand the leader of the Norwegian Women's Health Organisation was
      Elizabeth Kraft, whose husband, Captain Johan T. Kraft, a doctor in the army medical
      service, was a radical Liberal, attached to the local Rifle Ring. The deputy leader, Anna
      Stang, was married to the moderate Liberal Hans Georg Jacob Stang, member of govern-
      ment 1884, prime minister 1888–9 and MP 1892–4; mother of the Liberal George Stang,
      defence minister 1900–03, and a leading exponent of the active defence policy of the Lib-
      erals after 1895. Both Elizabeth Kraft and Anna Stang were active within the Norwegian
      Association for the Rights of Women (ibid., pp. 35–46; *Norsk Biografisk Leksikon*).

35   According to the *Norsk Biografisk Leksikon,* Thaulow occupied both of these positions, the
      first from 1881 to 1897, the other from 1889 to 1905. In the annual reports of the NDA,
      however, Major L. Chr. Dahl is mentioned as head of the Kristiania Circle in 1890.

36   Martin Sæter, *Over alle grenser: Norges Røde Kors 100 år* (Oslo 1965), p. 38, cited in Bringslid,
      'Norske Kvinners Sanitetsforening', p. 21.

37   Bringslid, ibid., pp. 21–5.

38 Ibid., pp. 51–2, 99–107; Agerholt, *Den Norske Kvinnebevegelses Historie*, pp. 196–9; for Qvam, see *Aschehougs Konversasjonsleksikon*.

39 Agerholt, *Den Norske Kvinnebevegelses Historie*, pp. 207–11.

40 *Nylænde*, 15 Oct. 1889, pp. 308–10.

41 Ibid., 1 Feb. 1890, p. 39.

42 Deborah Gorham, 'Vera Brittain, Flora MacDonald Denison and the Great War: the Failure of Non-Violence', in Ruth Roach Pierson (ed.), *Women and Peace: Theoretical, Historical and Practical Perspectives* (London 1987), pp. 137–48.

43 The following account is based on Ida Blom, 'The Struggle for Women's Suffrage in Norway, 1885–1913', in *Scandinavian Journal of History*, v, no. 1 (1980), pp. 3–22, and *idem*, 'Women's Politics and Women in Politics in Norway since the End of the 19th Century' in ibid., xii, no. 1 (1987), pp. 17–33 (reprinted in S. Jay Kleinberg [ed.], *Retrieving Women's History. Changing Perceptions of the Role of Women in Politics and Society* [Berg/Unesco 1988], pp. 254–77).

44 Anna Gjøstein was a member of the Stavanger branch of the Norwegian Association for the Rights of Women, chairing the branch in 1898–9. In 1901 she became the first leader of the Labour Women's Association in Stavanger (see Britt Marit Foldøy, 'Frivillig organisering blant kvinner i Stavanger fram till 1. verdenskrig' [unpublished MA thesis, Department of History, University of Bergen 1982], pp. 127, 199). She was married to Johan D. Gjøstein, a teacher who was a Labour Party MP, 1912–24 (*Norsk Biografisk Leksikon*).

45 Fuglum, *Norge i Støpeskjeen*, pp. 105–30.

46 Agerholt, *Den Norske Kvinnebevegelses Historie*, p. 218.

47 Blom, 'The struggle for women's suffrage'.

48 Johan Scharffenberg, in *Social-Demokraten*, 2 Aug. 1905. See also Blom, 'The Struggle for Women's Suffrage'.

49 Blom, ibid.

50 See, for instance, Berenice A. Carroll, 'Feminism and Pacifism: Historical and Theoretical Connections', in Roach Pierson, *Women and Peace*, pp. 2–28.

51 Dorothy Thompson, 'Women, Peace and History: Note for an Historical Overview', in ibid., pp. 29–43.

SOCIAL IDENTITY IN WAR: FRANCE, 1914–1918

John Horne

Action is social in so far as, by virtue of the subjective meaning attached to it by the acting individual or individuals, it takes account of the behaviour of others and is thereby oriented in its turn. (Max Weber)[1]

It is only too easy to think of the language and reality of twentieth-century 'total' wars as being those of organisation, since that was the understandable preoccupation of those who ran them. Yet without the *actions* of millions of individuals, such wars are inconceivable. And although that action often occurred under constraint (e.g. mass conscription), the two world wars could never have been fought had they relied on coercion. The state's preoccupation with public opinion and morale in both conflicts is the surest indication that the more or less willing commitment of the action and energy of the belligerent populations was vital for prosecuting the wars.

But social action – if we borrow Max Weber's definition – is not merely the external measurement of cause and effect. The essence of social action lies in its subjective significance for the individuals and groups undertaking it and also in the reciprocal influence of other individuals and groups in shaping such action. Society, from this angle, can be thought of as constituted by a tissue of mutual perceptions and value judgements by different groups, which coalesce into structuring principles of social consensus or diverge in social conflict. Action thus understood both proceeds from, and helps form, social identities. And the meaning and language of social morality (what is felt to be 'fair' or 'unjust', acceptable or unacceptable) is the medium in which the relation between social actors is worked out. What I wish to explore here, through the example of France and the Great War, is the proposition that the world wars generated particular forms of social action and identity in the sense which I have outlined, and that these constituted a fundamental dimension of the war experience, helping determine the degree of resilience or vulnerability of the nations involved. The discussion will be limited to the 'home front', which has received less attention in this regard than the fighting front.[2]

It might be argued that social action and identity in wartime are really subsumed in two other questions – the political meaning of the war, on the one hand, and the relationship of more permanent social formations and identities to the conflict on the other. Certainly, any question of identity and meaning in relation to what was, after all, a political event might itself be considered political. But the Great War posed a series of more narrowly political questions which in turn generated specifically political discourses (national unity, war aims, peace). These can be distinguished from the broader field of social action and identity which concerns us here, although there is no clear dividing line between the two.[3] The impact of the war on abiding social structures (such as gender or the working class, subjects of much of the work in this area) is clearly related to our enquiry.[4] But to remain within long-term categories of gender or social class, is to run the risk of missing (by assuming we know it in advance) a distinctive framework of wartime experience and action.

An exploration of the modes of action and identity stimulated by the war among civilians encompasses the models of wartime behaviour promoted officially or unofficially as part of the process of societal mobilisation, as well as responses which were either distinct from, or opposed to, these models. The sources for the former are to be found in official state pronouncements, in a variety of private bodies engaged in mobilising society, and in vehicles of cultural influence such as the press. Sources for more autonomous or refractory social action derive above all from the state's wartime apparatus of surveillance.[5]

The first year of the war was crucial for suggesting model responses to the conflict. The surprise outbreak of hostilities, the ubiquitous feeling that France was the victim of premeditated aggression, and the shock-waves triggered by the German invasion and near-capture of Paris in September, galvanised the energies of the nation. Normal values and assumptions seemed suspended. Exceptional qualities were required.

Perhaps inevitably, with the nation's fate pivoted on the Western front, the imagery of idealised behaviour was dominated by the demotic figure of the *poilu*, the hirsute soldier-hero of the Marne. Of course, physical selflessness by civilians did occur.

Invasion and occupation (as earlier and later) furnished occasions for exemplary behaviour by ordinary people. Resistance to the German occupation of the north-east and the stubborn survival of civilian life in the fighting zone itself, provided civilian examples of heroic behaviour. Gifts flooded in from all over France, for example, to the four primary schools which continued to operate in the champagne cellars under shattered Rheims.[6] But by and large, in a conflict which, unlike the Second World War, rarely found civilians in the front line, the ideal civilian comportment was subordinated to the supreme measure of combat experience, and was cast in a less heroic mode. As it emerged in 1914–15, this idea essentially displayed three attributes – duty, sacrifice, and solidarity.

For civilians, duty meant above all maintaining normality, so far as possible, in the absence of the soldiers – to garner the half-cut harvest of August 1914, to work the fields and reopen the schools the following autumn, and ultimately to man the munitions effort. Change was naturally required to sustain such 'normality' – women 'replacing' men in their peacetime functions, the old returning to work, the young shouldering burdens beyond their years. The language of heroism often consecrated this process, but with a shift in meaning, since the heroism in question became that of stoicism and endurance.

Examples are numerous. One of the principal agricultural associations issued diplomas and medals to heroic peasant women who had kept the family farm going under difficult circumstances. Thus the widow Balluais, of the Mayenne, mother of eight daughters and seven sons (six at the front) was cited for exemplary husbandry on her 43 hectare-holding with the aid of five of her daughters, while Mme Boude, with three children under thirteen, continued to farm her fifty hectares in the Marne (pillaged by the Germans in 1914) with only one labourer.[7] In 1915 the socialist architect of the munitions effort, Albert Thomas, rhetorically militarised the munitions workforce by comparing their heroism and sacrifice to that of the army, and in October, the leading weekly pictorial, *L'Illustration*, portrayed Thomas congratulating soldiers mobilised to the arsenal of St Chamond, in the Loire, who had won the *Croix de guerre* at the front – a fusion of military and civilian duty.[8]

Sacrifice was a theme closely allied to duty. The sacrifices extolled for civilians were partly material – the suspension of labour legislation, for example, in the interests of higher industrial output. But above all, sacrifice was defined in moral terms, as the supersession of the narrow self-interest appropriate to peacetime by a generosity of spirit and dignity of bearing to match the gravity of the hour and the supreme sacrifice which over four hundred thousand French soldiers had made by the end of 1914. Amongst upper-class *parisiennes* (according to one observer), dress was simplified, jewelry shunned, and 'the simple, laborious life was adopted everywhere'.[9] And when the theatres of Paris finally reopened in January 1915, they avoided boulevard comedies and mere 'distractions' in favour of patriotic recitations for the benefit of war charities.[10]

Solidarity, not only with the wounded, the disabled and the soldiers locked in the trenches, but also with refugees and other civil war victims, became the civilian substitute for action in combat. Nationally and locally, a host of charitable organisations sprang up connected to the war. Patriotic in the most general sense, this response was less political than a defensive reflex against the physical violation of the national social fabric – and its origins often lay in emergency responses to the August invasion. Thus, in Paris, tens of thousands of despondent Belgian refugees camping in the streets late in August provoked the mobilisation of charitable energies." In Rennes, the town crier announced the unexpected arrival of two thousand refugees with the result that as many meals were spontaneously delivered to the station in the middle of the night." Yet the front line soldier and the wounded remained the chief focus of civilian solidarity. Bourgeois women and their existing philanthropic bodies (such as the Croix Rouge, the Association des Dames Françaises, and the Union des Dames de France) played a key role in the charitable effort, and this in turn expressed the high moral tone which the war initially seemed to demand. But the reflex of solidarity spread across all classes. In Paris, the labour movement operated alongside the St Vincent de Paul, while the remotest rural primary schools turned into charitable workshops, knitting for the front."

In all, the idealised model of civilian comportment provided an initial image of cohesion which helped the home front, for all its diversity, establish a sense of identity in relation to the war. The supremacy of the soldier defined the officially-endorsed hierarchy of wartime social morality. Not only women (as some historians have pointed out) – in spite of their new-found functions – but the entire home front was morally subordinated to the male authority embodied in the warrior.'" Young and old were commonly portrayed as bound to acknowledge the almost mystical status of the returning soldier. 'Our brother in the trenches', wrote Henri Lavedan, the resident moralist of *L'Illustration*, 'though still our junior has assumed with his uniform the authority of one senior to us.' The primary school teachers' professional association in 1915 issued an illustrated booklet to show youngsters the respect due to wounded soldiers and military widows.'"

Sometimes class terminology was invoked to express this relationship – as with the union of classes in the munitions effort. But the commonest metaphors were those of gender and the family. Thus, the two emblematic figures of 1915 in the popular press were the wounded soldier and the nurse. The front cover of the biggest-circulation women's weekly, *Le Petit Echo de la Mode*, was dominated by women in soberly elegant nursing uniforms tending nomi-

nally wounded – and eminently presentable – officers. But both images in fact
permitted an exploration of the broader social morality linking front and rear.
The figure of the nurse encapsulated the complementary role of the rear in the
form of feminine ministration to masculine needs, while the wounded soldier
stated in extreme form the superior moral claim of the combatant. A full-page
drawing in *L'Illustration*, in February 1915, showed a wounded soldier, eyes ban-
daged and hand in sling, walking on his mother's arm and receiving the saluta-
tion of civilian society – a boy in sailor suit standing to attention, an *ancien* of
1870 saluting, the neighbours doffing their hats. The caption read:

> She hopes that he will recover completely ... But if this hope is not realised, if the cruel scars
> remain on this dear face, mother and son would find a precious consolation in the respectful
> admiration shown by France towards her heroes.[16]

Yet it was the family which best reflected – both literally and metaphori-
cally – the idealised civilian identity and its relationship with the front soldier. After all, the most general experience in 1914–15 was that of separation,
as mass call-up divided families and left a rear composed mainly of the young,
the elderly and women. Family solidarity became a focal image for civilian
comportment. The posters of the first war loan campaign in 1915 portrayed
stoic mothers with children waving off resolute soldier-fathers, and elderly
peasant couples, the shadowy image of their mobilised son floating over them,
emptying their life savings on to the war loan counter. The family, so the
message went, was the basic social unit in whose defence the war was being
fought, and the preservation of its moral unity by solicitude for the departed
soldier was the elementary duty of the civilian.

In some respects, this civilian ideal was based on the cultural attributes of
the bourgeoisie – on a characteristic sense of responsibility and social obligation, and a natural readiness to speak for the nation.[17] In conjunction with
patriotism, it provided a language with which the social élites could seek to
direct popular action.[18] Yet, it cannot be assumed that the model thus presented in 1914–15 was simply imposed on other social groups or failed to express
wider responses to the war. True, there is some evidence of indifference or
resistance. In remote or particularly Catholic rural areas, the authorities dis-
cerned a more traditional response, linked to a revival of popular religiosity,
in which the war was accepted in a spirit of resignation, and even perceived as
divine chastisement for the separation of church and state eight years earlier.[19]
In certain urban centres, including Paris, some elements of the working class
felt betrayed by labour's official support for the war and signalled reluctance
to integrate into the national effort by the use of traditional revolutionary
language.[20]

Yet much working class opinion seems to have followed the leadership of the labour movement in endorsing the war, and accepting some version of a wartime social morality based on sacrifice and solidarity. 'Nowhere in the Paris urban area', commented a police report of May 1915, 'has popular morale weakened to the point of losing confidence in the ultimate success of our arms or refusing the effort and sacrifice which are still needed'.[21] In the arsenal of Rennes, in 1915–16, the workers themselves suggested that the purchase of war bonds should be stopped out of their pay.[22] And the extremely low strike rate in 1915–16, though due to many factors, indicates a marked reluctance to betray solidarity with the front.[23] Of course, working class and peasant women who substituted for their husbands were usually driven by economic necessity, though even here the possibility of simultaneous patriotism should not be underestimated.[24] But it is clear that the effort of solidarity in 1914–15 was massive and popular in both urban and rural milieux. One Breton mayor in August 1914 (among many possible examples) proudly listed to the *Préfet* the contributions from his commune, including donations equivalent to a third of a daily wage by the bulk of peasant farmers.[25]

As the war turned into a question of endurance, however, the ideal civilian identity defined in the early months was exposed to two types of stress. The first came from the nature of the military conflict. Already in 1915, the reality of strategic deadlock and unprecedented casualty rates was evident, but by 1916–17, civilian opinion confronted the prospect of an apparently interminable, and possibly unwinnable, war of attrition. The second source of stress was the differentiation of social function and of economic condition which attended the mobilisation of a 'total' war effort. Together, these developments produced a more complex relationship between soldiers and civilians, and also clouded relations among civilians themselves. Over time, it was impossible to sustain a notion of civilian duty which subordinated every other consideration to the supreme sacrifice of the heroic soldier. 'Normalising' the war was ultimately incompatible with the suspended animation of daily life initially felt to be the most appropriate comportment. Upper-class circles were marked by uncertainty in 1915–16 over this question of appropriate behaviour. The Paris stage returned to comedy and light relief after a long debate as to whether 'the theatre in wartime should be grave or pleasant, instructive or amusing'. The conclusion was that civilian morale required diversion.[26] Fashion reawoke, and was justified in the same terms.[27] At a more fundamental level, the conflict emerged through the tension between the roles demanded of women, especially in the middle class. *Le Petit Echo de la Mode* urged its readers until the end of the war to play the role of loyal subordinate to the front soldier – preferring helpful advice on how to behave during a husband's home leave – with deference, sobriety and devotion. Yet the paper also recognised that women's new role as *remplaçantes* for men, including the temporary acquisition of legal rights as heads of the family, entailed novel manifestations of energy and independence.

'The longer the war lasts', the paper editorialised,' the more this situation develops which forces women to take on men's roles ... What could be less astonishing than to find their bearing and language changed: they adopt a resolute, independent air because they have the responsibility, and also the freedom, of being in charge ... ' [28]

These necessary shifts in civilian interpretations of wartime duty, and of the moral code binding the front to rear, were inevitably pregnant with possible misunderstanding and acrimony. There is a good deal of evidence that soldiers on leave were shocked by the more frenetic and debauched aspects of the big cities – night life, easy money and conspicuous consumption. As one officer (taken as an example by the postal control of Bourg-en-Bresse in late 1916) wrote of his leave in Paris: 'In a restaurant, I heard lots of talk of money by people who seemed to me to be really profiting from the war. They irritated me, and I found these people morally ugly. I am happy to be back amongst my men.' [29]

And the same postal control reported in early 1917 that 'this is the theme which one finds treated endlessly. The men endure with resignation ever-present danger, the rigours of winter ... But a constantly recurring leitmotiv ... is their hostility to war profiteers.' [30] This revulsion led to a discernible current of antipathy among soldiers to home front hypocrisy which masked profiteering with patriotism, and it generated the sentiment that a little civilian suffering (family always excepted) – such as deprivation and hunger – would be no bad thing. [31]

The greater independence of women, plus the increased possibilities for extra-marital affairs, in obvious ways weakened the reciprocal moral bond between soldier and civilian. Sexual jealousy emerged as one element in the attitude of the front soldiers, and of workers combed-out for the trenches in 1917–18. The latter feared in particular that their women as well as their work would be taken by immigrants (over half a million of whom were drafted in during the war). [32] Although the divorce-rate declined during the conflict, it rose from 1919, more than making up for the wartime trough and retrospectively suggesting that the war profoundly disturbed relationships. [33] The authorities were especially concerned at the opportunities provided by large-scale employment of women in munitions factories for more independent sexual behaviour.

Yet in other ways, the modification of civilian action and identity may well have reinforced, rather than weakened, the identification of the rear with the front. This is perhaps clearest with the notion of sacrifice. Sacrifice, which had initially been symbolic – a matter of bearing – rapidly assumed all too tangible a form through the loss of loved ones in combat. The need to recognise and endorse mourning and enrol it in the social morality of wartime was understood early on. From December 1914 there was a flurry of schemes to commemorate the fallen of each commune, with an acrimonious public debate in 1916 on the most appropriate way of doing this.[34] In 1915, the 'Cocarde du Souvenir' explicitly confronted the void caused by anonymous death when it undertook to place a named tricolour cockade on temporary graves so that the identity of the dead at the front should not disappear in the chaos of war. In September 1915 the first military monument was raised on the site of the 'miracle' of the Marne a year before. And from 1916, the government awarded a formal 'Diploma of Honour for Soldiers fallen for the Fatherland' to every afflicted family.[35]

Loss, in other words, lent substance to civilian sacrifice, and the government and press sought to endow this potentially fraught connection between front and rear with the correct patriotic significance through advice and official commemoration. Although it is difficult historically to probe the private significance of grief, the state surveillance machinery (prefects' reports and postal control) occasionally hints that loss strengthened the sense of commitment to the front soldier, and to the outright victory which alone would justify the mounting toll of sacrifice. The Bourg-en-Bresse postal control noted the importance of bereavement anniversaries for renewing commitment to the war, quoting the following letter (in late 1916): 'Two years have passed since our dear Edouard gave his life for the noblest cause. From the spot which he knew how to defend a victorious offensive is now being unleashed ... No sacrifice has been in vain, that is our consolation'.[36]

The relationship between civilian identity and the expected outcome of the war was not unvarying. As studies of public opinion have shown, morale fluctuated, especially in 1917, when the abortive Nivelle offensive in April appeared to demonstrate the implacability of the military deadlock, and events in Russia and Italy in October-November seemed to strengthen the strategic hand of the Central Powers.[37] This had its effect on the expression of civilian solidarity with the front. Although the charitable mobilisation continued throughout the war, the authorities took the annual war loan to be an important measure of civilian moral identification with the soldier and of confidence in the outcome of the struggle. The propaganda campaign was a paradigm of the effort by the state and social élites to translate an idealised civilian identity into action, with committees of local notables seeking to stump up subscriptions. But almost everywhere the 1917 results showed a dip. In the Tarn, for example, the Prefect considered that rural support in the most isolated, and least republican, areas tended to detachment from the national effort.[38]

Here, however, we perhaps need to distinguish between support for the military and political direction of the war and identification with the soldier. There is evidence that a small minority of soldiers dissuaded their families from supporting the war loan in the interests of hastening peace. Advice to leave land untilled was on occasions given by peasant soldiers to their families in the same spirit.[39] But this indicates the closeness of soldier-civilian ties, rather than the reverse. The message of solidarity, in other words, could be read in several ways. The small, overtly pacifist movement deployed the language of family solidarity to reverse official rhetoric, trying to persuade wives and parents to end the war, not continue it.[40]

Yet despite the relationship of distrust and hostility which some facets of the home front aroused in the soldiers, and despite a subcurrent of sexual distrust and jealously (whose precise importance is hard to evaluate), solidarity of civilians with soldiers at the most elemental level on which the war was experienced, that of the family, held firm. Although solidarity did not automatically operate in favour of the official view of the war (and in a minority of cases was clearly deviant from it), overall it supported the basic cause of national resistance. This is clearly demonstrated by the evidence of soldiers' correspondence from the postal control records. Time and again the importance of the marital and familial ties binding front and rear is affirmed, with peasant soldiers (well over half the front line troops) remaining closely involved in running the family farm.[41] Additional evidence comes from the tight surveillance which the authorities kept on soldiers' home leave in 1917–18. Except in the two periods of depressed morale in 1917, when civilians' and soldiers' discouragement mutually reinforced each other, furlough habitually had the opposite effect. Renewed contact with the intimate home front of the family (increased by the High Command after the mutinies of May-June 1917) tended to restore morale and mutual identification for both parties.[42]

The differentiation of function and economic fortune entailed by the complex mobilisation of the nation's economic and industrial resources subjected civilian identity to a different kind of tension. Inequalities developed in the way in which duty, sacrifice and solidarity with the front soldier were practised by various categories of civilian, in sharp contrast with the initial ideal. These inequalities can be explored through the negative social identities generated by the war – the 'profiteer', the 'speculator', and the *embusqué*, or 'shirker'.

The stock figure of the 'profiteer' was rooted in the reality of an industrial mobilisation dependent on private industry and on exceptional profit levels. This reality was by no means limited to large firms and heavy industry. Military contracts (including clothing and food processing) disseminated the benefit of state contracts over a wide array of small businesses.[43] By 1916–17, the phenomenon of conspicuous display and consumption by a stratum of wartime *nouveaux riches* provoked distinct antagonism from civilians as well as from soldiers on leave.

Criticism was not confined to workers. But organised labour articulated most clearly a hostility to war profits which slotted into its own traditional anti-capitalist demonology and which also protested against the breach of an idealised civilian identity couched by official rhetoric, as far as the munitions effort was concerned, in terms of cross-class co-operation. Initially, longer hours and raised output seem to have been accepted in a spirit of duty and solidarity with the front soldier. But soaring inflation and a bitter battle over piece-rates (which employers drove down in order to increase output) produced a change of mood in 1915–16, and this fastened on the symbol of the 'profiteer'. In mid-1916, the special police commissioner for Rouen described this growing disillusion in a region comprising 60 factories and 40,000 munitions workers.

If the worker is paid on piece-rates, at first sight his interest would seem to be to work as hard as possible. But experience has taught the worker that if he produces a lot, the employer will take advantage of this to lower the piece-rate, so that the worker will have to work harder for a wage which remains the same. In order to avoid this disagreeable consequence, it is in the worker's interest to limit his output ... This new attitude (by the worker) is resulting in a systematic decline in productivity, just when output should be intensified.[44]

From 1915 on, disillusionment took the form of constant denunciation of 'scandalous profits' in labour meetings, with figures quoted for the major companies like Renault, Citröen, Schneider and Gnome-et-Rhône. As Merrheim, national leader of the metalworkers, expressed it in 1917: 'Duty, sacrifice, national defence! Empty words for *patrons* who only want more gold, always more gold.'[45]

In order to justify what might appear to be the trade union movement's sectional defection from civilian identification with the war effort, the majority leaders of the Confédération Générale du Travail (CGT) adapted the language of 'sacrifice' to their own ends by identifying the working class, rather than the front soldier, as the butt of real inequality in the mobilisation process. This rhetoric allowed it, in the name of the war effort, to summon the state to re-establish a modicum of equity – and thus restore the scale of wartime social morality in labour's eyes – by raising wages and controlling employers, and ultimately by introducing sweeping social reforms. In this sense, the development of a critique of class inequality in the war effort was entirely compatible with continued support for the national cause.[46] The local leader of the metalworkers in Le Havre urged a large meeting in September 1917 to 'follow the example of their English comrades, who have adopted the following maxim: everything for the country, everything for the nation, but nothing for those who live off our sweat.'[47]

With the image of the 'profiteer', wartime social morality registered the particular tensions between labour and capital within the industrial mobilisation. The figure of the 'speculator' represented a more diffuse but in some ways more serious disaggregation of agreed norms for civilian behaviour. It also embodied an older tradition of urban moral protest, largely dormant since the mid-nineteenth century, which insisted on adequate food supplies at a 'just' price. As inflation developed along with limited shortages of certain goods (though never on the same scale as in the Central Powers – thanks to the Entente's access to world food supplies),[48] the explanation tended to be sought by the urban consumer in speculation (and therefore hoarding) by those traditional antagonists, the peasant and the shopkeeper.

In reality, the explanation lay at least as much as in the simple imbalance of supply and demand and also in the government's reluctance to control market forces. But a powerful mood of consumer discontent and protest emerged which, though expressed with particular force by organised labour, spread (in the eyes of the police) to wider elements of the urban population, including some of the *classes moyennes* (lower middle class). This consumer protest rediscovered traditional language and gestures. Markets were invaded and traders forced to sell at the 'just' price. The police commissioner of Carmaux (Tarn) described a typical incident in January 1916:

The miners and glass-workers of Carmaux went with their wives to the pig fair yesterday, in order to prevent hoarding. As soon as the army purchase agents appeared ... they were jeered and booed by the men and women. The demonstration went no further and no violence occurred, but faced with this attitude the agents deemed it prudent to quit the market without making any purchases (prices fell accordingly!).[49]

The Ministry of the Interior's survey of social movements and public opinion in 1917–18 identified these consumer issues month after month as the principal grievance. It was not just that urban consumers rediscovered the 'ever increasing greed of the merchants and the rural producers' but that this apparent greed directly contravened the equality of sacrifice, and thus the uniform moral yardstick of civilian action in relation to the war. The report for February 1918 concluded that 'speculation fuels growing doubt in the working class about the equality of sacrifice, a doubt which could have serious consequences'.[8]

For a traditional problem, there was a traditional answer — *taxation*, or price controls, and requisitioning, to prevent hoarding. In an explicit revival of the language of 1790s Jacobinism, the press and, politicians of the broad left demanded the application of this populist policy, and, as the Ministry of the Interior confirmed, the pressure for widespread rationing, to which the Clemenceau government acceded in 1918, came overwhelmingly from workers and other urban consumers.[9] But the point is that for much of the war, these economic antagonisms which set country against town, and producer and retailer against consumer, negated the image of the heroic peasant and the idea of a common civilian duty to the front soldier.

The issue of the *embusqué*, or shirker, reversed the terms of the wartime moral equation between civilians. Where the 'profiteer' and 'speculator' stood in an implicit binary relationship to a popular or working class victim, the negative image of the *embusqué* highlighted a different constellation of unequal wartime sacrifice. Although subject to a variety of meanings, the term as used by civilians denoted the differential toll of front line service, and therefore death, on families in different social situations. In part this was a question of age. As early as November 1914, families with fathers in their forties sent to the front condemned younger men mobilised to specialist functions in the rear.[10] But it was also a matter of specialisation of function — inevitable once a complex mobilisation of economic resources was under way — which breached the principle of equality of obligation to military, especially front line, service, and thus fractured the simple complementary model of the rapport between home and fighting fronts. The issues of age and function became hopelessly tangled with the chaotic recall to war factories of workers who had been mobilised at the front. This resulted in many younger workers finding sheltered work on munitions while older workers remained at the front, and many who were not genuinely irreplaceable skilled workers, or even workers at all, managing to avoid the risk of the ultimate sacrifice. *Embusqué* referred to any mobilised men who had succeeded in getting factory work without justification.[11]

But as the death-toll rose, the term easily came to denote the privileged status of the working class in general, whose families were sheltered from separation and the risk of death by comparison with those of the bulk of front line soldiers, able to claim no such exemption. As such, the campaign against *embusqués* waged by the press, politicians, and *ad hoc* interest groups expressed a barely-disguised vein of peasant and lower middle class hostility against urban workers, captured by a letter cited in late 1916 by the Bourg-en-Bresse postal control:

A large part of the population lives on the war while the other part suffers and dies from it; while the bulk of skilled workers are well sheltered in workshops and earn more money than in peace-time, while all the civil servants, most of them *embusqués*, continue after twenty-six months of war to get their salaries ... all the agriculturalists without exception, all the shopkeepers, will stay on the field of battle, and the tiny number who come back will be mutilated, ill and unable to work again.[54]

The government sought unsuccessfully to ensure that all younger mobilised men, unless absolutely irreplaceable in home front occupations, served in the trenches. The chaos involved in such rationalisation would have been counter-productive for the war effort as a whole, as those generals, politicians and civil servants most closely involved soon realised, with the result that official attempts to equalise the burden of death usually ended by being quietly shelved. The most the authorities could do to alleviate peasant bitterness in particular was to ensure generous home leave at harvest time, and to demobilise the oldest peasant soldiers in 1917–18.

Overall, the action of civilians, the framework of social morality within which this occurred, and the wartime identities to which it gave rise, form an important dimension of the French experience during the Great War. Social identity and action were not static. The simple model of behaviour and the subordinate relationship to the front line soldier which emerged in the early months were exposed to considerable strain and modification under the twin impact of a war of attrition and a complex mobilisation of civilian energies in their own right. But the role of wartime social morality as a regulator of the images of self and other among different social groups, and of relative contributions to the national effort, remained central.

Naturally, the components of wartime social identity drew on longer-term social formations and traditions. Yet they were equally marked by relationships which were quite specific to the war, such as that of the front line soldier risking his life to the rest of the population, or of national survival (rather than straight profitability) to industrial production. For this reason, 'sacrifice', even more than duty or solidarity, became the central value of wartime morality, since it encapsulated the extraordinary nature of the enterprise while defining the various contributions to the effort and their (often conflicting) claims to recognition and consideration. In this process, the family emerges as being especially important – since it was the divided family which most ubiquitously expressed the relationship of civilian to soldier, and since the full, differential impact of the ultimate sacrifice by front line soldiers was reflected back into civilian society by the widely varying vulnerability of families from different social groups to bereavement and resultant economic hardship.

The state's preoccupation with the issue provides further evidence of the importance of civilian social identity and morality during the war. The authorities sought to do more than monitor public opinion and morale. They tried (within limits and subject to conflicting political pressures) to regulate sensitive questions of wartime social morality, such as *embusqués*, war profits, and rationing, as well as articulating ideal models of behaviour which would help strengthen wartime social cohesion.

Ultimately, a judgment as to France's resilience or instability in the face of the Great War can only be made in a wider context including other factors, such as military strategy, living standards, and more overtly political attitudes to the war. But the shaping impact of a specifically wartime social morality needs to be evaluated as part of the calculation. In many respects, the various perceptions of unequal sacrifice by different groups cancelled each other out, or at least mutually modified each other, and thus contributed in a negative way to the social cohesion of wartime France. This is particularly clear in relation to the perceived acceptability of wartime strikes. In May 1918 the most serious strikes of the war occurred when pacifist-motivated labour activists resisted the 'combing-out' of workers for the trenches and sought to paralyse munitions production during the German offensive in order to force a negotiated end to the conflict. In effect, they rejected the prevailing terms of civilian identification with the front line soldier while benefiting in the eyes of many from a protected, privileged, position. Their resultant isolation from public opinion at large, including many front line soldiers, contributed to the movement's defeat.[5] Conversely, the movement revealed the importance still attached to the ideal of social solidarity between home and fighting fronts. The 1918 strikes thus doubly underline the importance of social action and its resultant moral imperatives in explaining the comportment of a society faced with the test of a 'total' war.

NOTES

1   *The Theory of Social and Economic Organisation* (New York 1964), p. 88.

2   A. Cochet, 'L'Opinion et le moral des soldats en 1916 d'après les archives du contrôle

Cover of booklet produced in 1915 for primary schoolchildren
(Bibliothek für Zeitgeschichte, Stuttgart)

Lucien Jonas, 'Le salut au blessé', L'Illustration, 20 February 1915

postal' (*troisième cycle* thesis, University of Paris X-Nanterre 1985); S. Audoin-Rouzeau, *14–18: Les combattants des tranchées* (Paris 1986).

3 For the study of the political meanings of the war and of the related question of public opinion, see the works of J.-J. Becker, in particular, *1914: Comment les français sont entrés dans la guerre* (Paris 1977) and *The Great War and the French People* (1980; English trans., Leamington Spa 1985).

4 For gender, see J. F. McMillan, *Housewife or Harlot: the Place of Women in French Society 1870–1940* (Brighton 1981) and F. Thébaud, *La Femme au temps de la guerre de 14* (Paris 1986). For recent relevant examples of labour and working class history, see K. Amdur, *Syndicalist Legacy: Trade Unions and Politics in Two French Cities in the Era of World War I* (Urbana, Ill. 1986); P. Fridenson, 'The Impact of the First World War on French Workers', in R. Wall & J. Winter (eds), *The Upheaval of War. Family: Work and Welfare in Europe 1914–1918* (Cambridge 1988), pp. 235–48; J.-L. Robert, 'Ouvriers et mouvements ouvriers parisiens pendant la grande guerre et l'après-guerre. Histoire et anthropologie' (Doctorat d'Etat thesis, University of Paris-I 1989); and J. Horne, *Labour at War. France and Britain, 1914–1918* (Oxford 1991). No comparable attention has been paid to the impact of the war on rural society, the *bourgeoisie*, or the *classes moyennes*.

5 Those which have been consulted are the records of the postal control commissions which monitored the opinion expressed in mail leaving France (Service Historique de l'Armée de Terre [SHAT], 7N 979–1001, esp. the commission of Bourg-en-Bresse, 7N 987–8); reports by prefects on opinion in their *départements* in Archives Nationales (AN), F7 12970–13023 (uneven), SHAT 16N 1538 (June and September 1917), and in selected departmental archives; monthly synopses from August 1917 to August 1919 compiled on a national basis by the Second Bureau of the High Command from reports supplied principally by the generals commanding the twenty internal French military regions (SHAT 5N 268 & 6N 147); and monthly reports of the Direction de la Sûreté Générale (Ministry of the Interior), from Oct. 1917 to Oct. 1918 (SHAT 16N 1538 & Bibliothèque de Documentation Internationale Contemporaine (BDIC), F 43 Res. 1).

6 S. Trouvé-Finding, 'French State Primary Teachers during the First World War and the 1920s: their Evolving Role in the Third Republic' (PhD thesis, University of Sussex 1987), p. 136. See also *L'Illustration*, 10 July 1915, for schoolchildren on the Lorraine front. For women as heroes in occupied France and the battle zone, see L. Abensour, *Les vaillantes. héroïnes, martyres, remplaçantes* (Paris: Chapelot 1917), pp. 137–218.

7 J.-H. Ricard, *L'Appel de la terre* (Paris: Payot 1919), p. 141.

8 *L'Illustration*, 4 & 25 Sept. 1915.

9 *Paris charitable pendant la guerre*, (Paris: Office Central des Oeuvres de Bienfaisance 1915), preface by R. Vallery-Radot, p. viii.

10 *Revue des Deux Mondes*, 15 Jan. 1915, p. 444; Archives de la Préfecture de Police (Paris), B/a 1614, report of Nov. 1915 on the capital's theatres and cinemas since the beginning of the war.

11 *Paris charitable pendant la guerre*, p. ix.

12 *Les Nouvelles Rennaises*, 8 Oct. 1914.

13 *Paris charitable pendant la guerre*; Horne, *Labour at War*, pp. 84–9; S. Trouvé-Finding, 'French State Primary Teachers', pp. 117–39.

14 M. & P. Higonnet, 'The Double Helix', in M. Higonnet, J. Jenson et al., *Behind the Lines. Gender and the Two World Wars* (New Haven 1987), pp. 31–47.

15 H. Lavedan, 'La maturité de la guerre', *L'Illustration*, 10 Apr. 1915; G. Roson, *Aux enfants de*

France. Un vieux mobilisé (Paris: Fédération des Amicales d'Institutrices et d'Instituteurs de France et des Colonies 1915).

16   Lucien Jonas, 'Le salut au blessé', *L'Illustration*, 20 Feb. 1915.

17   A. Daumard, *Les bourgeois et la bourgeoisie en France* (Paris 1987), pp. 244–5.

18   For the role of the local élites in framing popular mobilisation, see A. Jacobzone, *14–18. En Anjou, loin du front* (Maine-et-Loire: Ivan Davy 1988), pp. 129–36.

19   Archives départementales (AD), Tarn, 10R 1/2, rep. of Commissaire de Police, Mazamet, to the Sous-Préfet of Castres, 17 July 1915; AD Ille-et-Vilaine, 3M 340, rep. of Préfet to the Minister of the Interior, 29 May 1915.

20   AN F7 13574, rep. 27 Aug. 1915.

21   AN F7 13574, rep. 7 May 1915.

22   AD Ille-et-Vilaine, 4M (uncatalogued), Main-d'oeuvre de la défense nationale, rep. 22 June 1917.

23   The number of striking workers fell from 161,000 in 1914 to 9,000 in 1915, and 41,000 in 1916, before soaring to 436,000 in 1917 and 386,000 in 1918 (Horne, *Labour at War*, p. 396).

24   Eg. the oral testimony of Juliette Eychenne in R. Cazals *et al.*, *Années cruelles 1914–1918* (Aude: Atelier du Gué 1983), p. 60.

25   AD Ille-et-Vilaine R (uncatalogued), oeuvres de guerre, 1914–15.

26   *Revue des Deux Mondes*, 'Revue dramatique', 15 Mar.1916, p. 434; Archives de la Préfecture de Police (Paris), B/a 1614, report of Nov. 1915, which considered that by the autumn of 1915, light relief was the predominant motive of theatre-goers, though 'patriotic tableaux' were often interspersed with comic numbers in variety shows.

27   Abensour, *Les vaillantes*, pp. 24–5.

28   *Le Petit Echo de la Mode*, 28 May 1916.

29   SHAT 7N 988, Bourg-en-Bresse commission, rep. Nov.-Dec. 1916, p. 14.

30   Ibid., Jan.-Feb. 1917, p. 6.

31   Cochet, 'L'Opinion et le moral', ii, pp. 331–3.

32   Ibid., pp. 325–7; G. Baconnier, A. Minet, L. Soler, *La plume au fusil. Les poilus du Midi à travers leur correspondance* (Toulouse 1985), p. 306; J. Horne, 'Immigrant Workers in France during World War I', *French Historical Studies*, 14/1, 1985, pp. 57–88.

33   Li Mon, *Le divorce en France* (Paris 1936), p. 86.

34   J. Ajalbert, *Comment glorifier les morts pour la patrie?* (Paris: Crès 1916).

35   AD Tarn 10R 3/1, poster of 'La Cocarde du Souvenir', 1915; *L'Illustration*, 18 Sept. 1915, for commemoration of the Marne; AD Ille-et-Vilaine 6Z 48, for the Diploma of Honour.

36   SHAT 7N 988, rep. Sept.-Oct. 1916, p. 13; Cochet, 'L'Opinion et le moral', ii, pp. 263–5; P.J. Flood, *France 1914–18. Public Opinion and the War Effort* (London: Macmillan 1990), p. 180.

37   P. Renouvin, 'L'Opinion publique et la guerre en 1917', *Revue d'histoire moderne et contemporaine*, Jan.-March 1968, pp. 4–23; Becker, *The Great War an the French People*, pp. 217–48.

38   P. Vatin, 'Politique et publicité. La propagande pour l'emprunt en France, 1915–1920', *Revue d'histoire moderne et contemporaine*, April-June 1980, pp. 207–36; AD Tarn 10 R 5, and for corroboration from the Isère of the dip in war loan subscriptions in 1917, especially in the countryside, see Flood, *France 1914–18*, pp. 161–2.

39   Vatin, 'Politique et publicité', pp. 217–8, 222; SHAT 16N 1538, Ministry of the Interior's enquiry among the prefects into the state of morale in each département, June 1917, replies for the Cher, Loire, and Loir-et-Cher.

40   See examples in AN F7 13372, 'La propagande pacifiste', a lengthy police compilation of

pacifist statements, such as the 'Manifeste du Comité Lyonnais pour la Reprise des Relations Internationales', Oct. 1916, which appealed 'To women, who have sensitive hearts, who are made to procreate and love, do you see the horror of the battlefields, can you not hear the cries of suffering of your sons, your husbands, your brothers....?' (appendix no. 5).

41 Cochet, 'L'Opinion et le moral', i, pp. 174–6; Baconnier, Minet and Soler, *La plume au fusil*, pp. 291–4.

42 SHAT 5N 268, reports of the generals commanding the internal military regions. The negative effect on home opinion of soldiers on leave in May-June 1917 is reported on by the prefects in the Ministry of the Interior's enquiry in June 1917 (SHAT 16N 1538).

43 For the operative principles of the industrial mobilisation, see M. Fine, 'Guerre et réformisme en France, 1914–1918', in L. Murard and P. Zylberman (eds), 'Le soldat du travail', *Recherches*, 32/3, Sept. 1978, pp. 305–24; A. Hennebicque, 'Albert Thomas et le régime des usines de guerre, 1915–1917', in P. Fridenson (ed), *1914–1918: L'Autre front* (Editions ouvrières 1977), pp. 111–44; G. Hardach, 'La mobilisation industrielle en 1914–1918: production, planification et idéologie', in ibid., pp. 81–109; and Horne, *Labour at War*, pp. 143–8. For the wide dissemination of state contracts in the Isère, see Flood, *France 1914–18*, pp. 137–9, corroborated for another département by AD Ille-et-Vilaine, 4M, list of factories working for the national defence in 1918.

44 AN F7 13367, rep. 22 June 1916.

45 AN F7 13366, rep. of meeting of the Comité Intersyndical des Métaux of the Seine (Paris region), 20 Feb. 1917.

46 Horne, *Labour at War*, pp. 265–7.

47 Rep. in AN F7 13367.

48 A. Offner, *The First World War. An Agrarian Interpretation* (Oxford 1989).

49 AD Tarn 10 R 1/2, rep. 15 Jan. 1916.

50 BDIC F 43 Res. 1, rep. Feb. 1918.

51 BDIC F 43 Res. 1, reps. of Jan. and Feb. 1918. For economic Jacobinism, see Horne, *Labour at War*, pp. 92–3.

52 Eg. AN F7 12937, rep. of prefect of the Charente, 10 Nov. 1914, on the departure of men of the 94th Territorial Regiment (i.e. men in their later thirties and forties), which produced protests on this issue. See also the lead article by Gustave Hervé in his populist left-wing, but pro-war, paper, *La Guerre Sociale*, on 24 Nov. 1914, which condemned what he saw as a wave of popular '*embuscomanie*'. In Rouen, the first protest against shirkers, in the sense of young men who remained in desk jobs, came as early as 11 August in an article by the local deputy in the *Dépêche de Rouen*.

53 J. Horne, 'L'Impôt du Sang: Republican Rhetoric and Industrial Warfare in France, 1914–18', *Social History*, 14/2, May 1989, pp. 201–23.

54 SHAT 7N 988, rep. Oct.-Nov. 1916, p. 10.

55 For a summary of recent work on the strikes, see Horne, *Labour at War*, pp. 180–7.

Keith Jeffery

In examining this subject, in which I have concentrated on the period up to 1939, I have sought to investigate both the ways in which the Great War has been remembered in Ireland and the ramifications of that memory in cultural, social and political terms. I have been inspired, obviously, by Paul Fussell's brilliant and seminal study, *The Great War and Modern Memory*,[1] and Samuel Hynes' more recent *A War Imagined: the First World War and English Culture*.[2] Although the Great War was of cataclysmic and central significance in the development of modern Europe, little serious scholarship has been published on specifically Irish matters, apart from the excellent, pioneering work of the Trinity History Workshop in David Fitzpatrick (ed.), *Ireland and the First World War*.[3] Yet there is a wealth of material available which may not merely elucidate the specific impact of the war on Ireland, but might also help us towards an understanding of both the certainties and the ambiguities of Ireland and Irish people, north and south, in the first part of the twentieth century. This essay, however, should be regarded as no more than a preliminary study of some aspects of the topic and a contribution to what I hope will become a major area of Irish historical study.

I

This is not the place – though the problem is not at all irrelevant – to investigate in any detail what we actually understand by 'Irish' culture. What, for example, makes an Irish artist or composer? We ought, however, to recognise that the question has profound political implications. Reflecting on the frequent debates on this very subject, and the arguments about Irish and Anglo-Irish culture in the *Irish Statesman* during the 1920s, Terence Brown observed that they were 'disguised debates about politics. The underlying political issues were who should shape the new Ireland and what traditions, if any, should be predominant.'[4]

Music is especially difficult to evaluate in cultural terms because the gulf

between popular and high cultures (at this time at any rate) was comparatively greater than in either literature or visual art. What, in any case, are we to
make of a man like Sir Charles Villiers Stanford, born in Dublin (1852), educated at Dublin and Cambridge, Professor of Music at Cambridge University
and the Royal Academy of Music, an – perhaps *the* – establishment figure in
English music, who wrote an 'Irish Symphony' (number 3, 1887) and six celebrated Irish rhapsodies, yet also set music to the poems of that most patrioti-
cally English of writers, Sir Henry Newbolt, and in 1918 set a song called
'Wales for Ever'?[5] He seems more of a jobbing national composer than a
specifically Irish musician.

In 1915 Stanford, reflecting on 'music and the war', and the influence which
international war had had in 'awakening the highest forces of musical art',
asserted that 'at no time has a great country failed to produce great composers
when its resources have been put to the supreme test of war'.[6] The experience
of English (or, indeed, 'Irish') music in the Great War scarcely bears out this
confident assertion. Stanford himself, who was nearing the end of his musical
career (he died in 1924), composed very little during the war. He set verse by
Newbolt – 'The King's Highway' (1914) – and the veteran Etonian A.C.
Ainger – 'The Aviator's Hymn' (1917) – and dedicated some chamber music
to wartime themes. A choral piece, *The Last Post* (words by W.E. Henley),
which he had written during the South African War in 1900, was revived in
1916. It was performed in Belfast in October 1916,[7] but not since and none of
the other music has survived in the repertoire.

Stanford's younger compatriot, Hamilton Harty, born in Hillsborough
(Co. Down) in 1879, followed a similar career pattern, culminating as conduc-
tor of the Hallé Orchestra and a member of the English musical establishment. While he composed more explicitly 'Irish' music than Stanford –
including his tone-poem 'With the Wild Geese' (1910) and an Irish Symphony first performed in Dublin in 1904, but subsequently heavily revised in 1915
and 1924 – in the end his status as an 'Irish' composer owed more to his place
of birth than any intrinsic feature of his music. Harty, who spent two years
with the Royal Naval Volunteer Reserve on 'hydrophone duties' in the North
Sea,[8] did not write any specifically 'war' music, but the conflict certainly influenced the musical public's perception of his work. In November 1919, Harty
himself conducted a performance of his cantata *The Mystic Trumpeter* in Belfast.
This was a setting, completed in 1913, of words by the American Quaker and
pacifist poet Walt Whitman, which tells how the trumpeter's music evokes
the chivalry and martial spirit of crusaders, the power of love and the triumph
of joy in the future. In the original poem the trumpeter also conjures up war's

'wild alarums', but Harty did not set this stanza.[9] Nevertheless, the music critic of the *Belfast News-Letter* clearly interpreted the work in the light of the Great War, and it was hardly coincidental that the performance was scheduled for November 1919. 'The composer ... ', he wrote, 'has made his music eloquently expressive of the passion and pathos which were felt by Whitman as he thought of the horrors of war and proclaimed the power of love.'[10]

The war was not entirely forgotten in the South. In 1920 John F. Larchet, the leading Irish composer working in Ireland, Music Director at the Abbey Theatre and Professor of Music at University College Dublin,[11] published *The Legend of Lough Rea*. This piece for unaccompanied chorus retells the story of how once in seven years a black coffin may be seen on the waters of Lough Rea, a sign thought to herald the approach of pestilence and mortality. Death shall sweep over the land and 'the light laugh of mirth shall be changed into wailing,/The living shall weep for the dead of Lough Rea'. 'Ah! Youth of the Land', it ends, 'Death! Death!'[12] It is, I think, not altogether fanciful to put this lament in the context of a widespread post-war feeling that the war had been some sort of pestilence visited specifically on the 'youth of the land', and for the Irish nationalist (such as Larchet) the pain was increased by the recognition soon after the war that the sacrifice had been in vain. None of the promises with which John Redmond and others set the National Volunteers off in 1914 had been fulfilled.

To a much greater extent than Larchet, Stanford and Harty exemplify one type of Irish culture: a hybrid ascendancy culture which attempted to fuse an Irish heritage with broader British and European trends, in the case of music particularly with the powerful weight of German romanticism. The product of this effort – Harty's music after the Great War – was stranded in a kind of cultural limbo. Two developments had occurred: the Irish side had moved on and the political legacy of the war had led to a purer, more Gaelic definition of Irish culture (as, indeed, represented by Larchet), which left little room for Harty's Hillsborough Hibernicism. European music had moved on, too. The 'shock of the new' left composers of Harty, or Stanford's ilk far behind. Harty's Irish Symphony well illustrates this point. Much of it is based on Irish folksong, and as Henry G. Ley observed, 'the very fact of its euphony' led it to be seen as 'speaking a dead language'.[13]

In *Rites of Spring* Modris Eksteins has reminded us how colossal a role the Great War played in 'the birth of the modern age'. He took his title, and begins his book, with that 'landmark of modernism', Stravinsky's 'The Rite of Spring', which was first performed in Paris in May 1913.[14] In musical terms, however, both Stravinsky and the Great War passed Ireland by. Even Larchet, undoubtedly more 'Irish' than the other two, was scarcely much more radical. Indeed, it can be argued that his own ambition to help 'create a school of composers which would be truly evocative of the Irish spirit'[15] actually had a constricting and retarding impact on the development of new music in Ireland.

The pieces we have discussed have largely been forgotten, though local 'patriotism' prompted a few subsequent performances in Belfast of Harty's 'Mystic Trumpeter'.[16] It may well be that the real Irish musical legacy (if any) of the Great War is to be found elsewhere, perhaps in more popular culture. A completely uncharted, though potentially very rewarding, area would be to look at war and anti-war popular songs. Although the war may not have changed popular culture to any great extent, surely the vital indigenous Irish musical tradition responded to some aspects of the conflict. The single most famous popular song of the Great War has, of course, an Irish angle – 'Tipperary'. It was apparently written in 1908 by two American men, though Alice Smith B. Jay has claimed responsibility for the chorus.[17] The song, in common with other popular wartime lyrics, depicts stage Irishness: the simple, lovable Paddy who is not very bright, but a good fellow all the same.[18]

In so far as there is a popular musical conception of the Great War it exists in these soldiers' ditties – the 'Songs that Won the War'.[19] It was Joan Littlewood's genius to tap into this with *O What a Lovely War* in 1963. But the popular perception had existed for a long time. When R.C. Sherriff's classic and immensely successful Great War play, *Journey's End*, came to Belfast during March 1930, in the intervals the Grand Opera House orchestra played 'popular musical comedy melodies from 1916–17–18' and 'Tommy's War-time Marching Songs', including 'Tipperary' and 'God Bless King George'.[20] No doubt this was precisely what the audience expected and it accorded well with the sentimental realism of the play. As Samuel Hynes has noted, it contains all the basic elements of the Great War 'Myth': stock characters – a hard-drinking commander, innocent new lieutenant and so on – behaving within an English public school ethos.[21]

Journey's End opened in the Savoy Theatre in London in January 1929 and played for six hundred performances. In the late autumn of the same year Sean O'Casey's *The Silver Tassie*, which had been turned down by the Abbey Theatre in Dublin,[22] played for a few weeks at the Apollo Theatre. O'Casey himself remarked that the play 'was so different from the false effrontery of Sherriff's *Journey's End*, which made of war a pleasant thing to see and feel, ... all the mighty, bloodied vulgarity of war foreshortened into a petty, pleasing picture'. *The Silver Tassie*, by contrast, 'tried to go into the heart of war, and, to many people whom it blasted with dismay, it succeeded.'²³ The play tells the story of three footballing pals who return much changed from the war: one is blinded; one terribly crippled; and one is a hero with a VC. Here is the appalling impact of war, and in the Apollo Theatre production the second act, set 'somewhere in France', with hugely dramatic scenery designed by Augustus John, dominated by a great Howitzer, was much admired for its fierce power.²⁴

The Silver Tassie exemplified one of the two main themes in Irish war literature: the utter degradation and demoralisation of war, a theme which perhaps sprang from the author's concern with the human cost of the conflict. Liam O'Flaherty's 1929 novel, *The Return of the Brute*, fits firmly into this category. Described by the author of the standard work on O'Flaherty's novels as 'one of the worst [novels] ever published',²⁵ and by another critic as 'the most degraded fictional picture of the Great War written by a combatant on the British side',²⁶ the book is, however, yet admired by *aficionados* of war fiction for its raw power. O'Flaherty had himself served with the Irish Guards on the Western Front, had been severely wounded in 1917 and invalided home with what was described as *melancholia acuta*.²⁷ His novel is a tale of nine superstitious infantrymen at the Front during 1917 – 'doomed men waiting for death'.²⁸ Most of them die in shocking circumstances: one goes mad, kills the non-commissioned officer and runs in front of the enemy machine guns, another drowns in mud and a third is fatally lacerated while defaecating.

The other main literary theme is one of distance, disengagement or detachment from the war, exemplified above all by Yeats' very well-known airman from Kiltartan Cross: 'Those that I fight I do not hate,/Those that I guard I do not love.' The pilot did not enlist from some patriotic or ethical motive: 'Nor law, not duty bade me fight,/Nor public men, nor cheering crowds./A lonely impulse of delight/Drove to this tumult in the clouds.'²⁹ While this is a persuasive and attractive explanation, in 1919, when the collection *Wild Swans at Coole* was published, the absence of political engagement with either the issues or the nations involved in the conflict could also be regarded as a convenient *post hoc* rationalisation for the Irishman who had served in the British Forces. Yeats, too, in sharp contrast to O'Casey and O'Flaherty, was reluctant to dwell on the 'realities' of the war. 'If war is necessary, or necessary in our time and place,' he asserted in 1936, 'it is best to forget its suffering as we do the discomfort of fever, remembering our comfort at midnight when our temperature fell.' Arguing that 'passive suffering is not a theme for poetry', moreover, he omitted from his *Oxford Book of Modern Verse* much of what is now regarded as quintessential Great War poetry.³⁰

There is also an element of the old 'fighting Irish' stereotype here, if in a rather refined form. Bernard Shaw made the same point in typically mischievious fashion in the 1919 preface to his 'recruiting pamphlet' play *O'Flaherty V.C.* (1917). There was, he argued, no point turning to 'Irish patriotism to get men to enlist', rather one should

> appeal to his discomfort, his deadly boredom, his thwarted curiosity and desire for change and adventure, and, to escape from Ireland, he will go abroad to risk his life for France, for the Papal States, for secession in America, and even, if no better may be, for England.[31]

The absence of a political motive underlying service in the Great War occurs in Tom Kettle's most famous war poem, and although he asserted a Christian impulse, his telling phrase 'Died not for flag, nor King, nor Emperor', certainly touched a nerve in the post-war Irish world. It also inevitably echoed the Citizen Army's celebrated slogan, 'We serve neither King nor Kaiser, but Ireland'.

The two themes identified: the human cost of the conflict and detachment from the issues crop up again in the visual representation of the war. So, too, does the question of serving Ireland, or at least not serving England. In the early spring of 1916 William Orpen's pupil and studio assistant Sean Keating knew he would have to leave London and return to Ireland in order to avoid being conscripted. He tried to persuade Orpen to accompany him: 'I said to him before I went: "Come back with me to Ireland. This war may never end. All that we know of civilisation is done for ... I am going to Aran ... Leave all this. *You* don't believe in it.' But Orpen remained in England, claiming that everything he had he owed to England. 'This is their war', he said, 'and I have enlisted. I won't fight, but I'll do what I can.'[32] Orpen went to France to paint 'Dead Germans in a Trench', while Keating went to Aran to paint 'Men of the West', and later 'Men of the South' (itself a 'war' painting). Although Orpen never properly returned to Ireland after August 1915 (a one-day visit in 1918 was the only time he was in Ireland again before his death in 1931), he never lost his Irishness, at least in others' perception of him. Despite his unassailable position as the court painter to the British establishment, and as an official War Artist employed by the British Ministry of Information, he remained an outsider looking in. On the Western Front he was an 'Onlooker in France', the title he took for his illustrated war diary which was first published in 1921.[33]

There are two principal features about Orpen the Irishman and war artist. First was his detachment from the actual conflict, primarily the result of his status as an official observer of the war (which, for example, kept him from the front line), but also as an Irishman, not wholly engaged in what we might call the Anglo-German conflict. This detachment, however, and this is the second point, is coupled with an intense sympathy for the common soldier. In the preface to *An Onlooker in France* Orpen noted his 'sincere thanks for the wonderful opportunity that was given me to look on and see the fighting man, and to learn to revere and worship him.'[34]

We can see this illustrated most vividly in the third of his three great peace paintings, which he was commissioned to produce in 1919 for a total fee of £6,000.[35] In the first two, 'A Peace Conference at the Quai d'Orsay' and 'The Signing of Peace in the Hall of Mirrors, Versailles, 28th June 1919', the statesmen are dwarfed by the overwhelming size of the state rooms they occupy. There is not much sympathy here for the important individuals depicted: President Wilson, Clemenceau, Lloyd George and so on. But the third painting, 'To the Unknown Soldier in France', is rather different. It was originally intended to show some forty 'politicians and generals and admirals who had won the war', and Orpen conceived a similarly monumental setting in the 'Hall of Peace' on the way in to the Hall of Mirrors. But he abandoned this plan – 'And then, you know', he recounted, 'I couldn't go on. It all seemed so unimportant somehow beside the reality as I had seen it and felt it when I was working with the armies. In spite of all these eminent men, I kept thinking of the soldiers who remain in France for ever ... So I painted all the statesmen and commanders out.'[36] Orpen then painted a single coffin shrouded in a Union Flag, and placed ghostly semi-naked British soldiers on each side, rather like armorial supporters, the whole completed with two cherubs floating above holding green and gold garlands. 'The gilded pomp of the Palace of Versailles ... ', went one interpretation, 'is imaginatively contrasted with the ragged misery of the ghostly boy-soldiers who watch over the coffin of their comrade. Festooned Cupids and the Cross shining in the distance are symbols of the "Greater Love" of those who have laid down their lives.'[37] The picture reflected Orpen's own view that the only tangible result of the war had been 'the ragged unemployed soldier and the Dead'.[38]

The critics generally did not like it, although the public did, voting it 'Picture of the Year' at the Royal Academy's 1923 summer exhibition. The Imperial War Museum agreed with the critics and refused to accept it. Opinions divided on political lines. The *Daily Herald* welcomed it as 'a magnificent allegorical tribute to the men who really won the war'. The *Patriot* rather disagreed and called it

a joke – and a bad joke at that. Sir William Orpen, by the way is an Irishman, like his brother artist Sir John Lavery ... The English people are very patient and very indulgent. But perhaps they are not quite so stupid as some of the Irish who live amongst us suppose. When an Irishman who accepts the hospitality of this country and profits by its wealth and its culture espouses the cause of Sinn Fein like Sir John Lavery, or takes liberties with our feelings of reverence like Sir William Orpen, there is an obvious comment which rises to the lips, but is not usually uttered. If these are their feelings, why do they not go and live in their own country? Certainly to artists who have a turn for the grotesque and the fantastic desolation of war, there are excellent subjects ready to hand in Southern Ireland.[39]

Five years later Orpen offered to paint out the soldiers, cherubs and garlands and dedicate the picture as a memorial to Earl Haig. The Museum accepted this proposal.[40]

So Orpen shifted from the second characteristic of his war art – his human – humane – sympathy, albeit expressed in a strikingly (though not for him uncharacteristically) surreal fashion – back to a more detached and abstracted memory. The people are literally painted out. The history of this painting – itself a war memorial – precisely reflects a central choice faced by those commissioning the more familiar type of public war memorial. Do you choose for your design some figurative representation – a soldier (usually), civilian or symbolic figure expressing victory, mourning, or whatever? Or do you opt for an abstract structure, such as a column, obelisk, cross, or most famous of all, a cenotaph?

The second great Irish official War Artist was the Belfast-born John Lavery. Although, like Orpen, he was also an immensely successful portrait painter, his war paintings contain little of the human passion displayed by his Dublin-born compatriot. Most of his pictures depict the Home Front, principally in a conventional landscape mode. Painting warships far from the fighting, at Scapa Flow, Lavery conceded that he 'felt nothing of the stark reality, losing sight of my fellow men being blown to pieces in submarines or slowly choking to death in mud. I saw only new beauties of colour and design as seen from above.'[41] A similar pleasure with shape and form can be seen in his paintings of British airships, but they are not pictures of much passion, and Lavery himself afterwards dismissed his war paintings as 'dull as ditchwater'.[42]

Images of the Great War were also being produced by William Conor in Belfast. Of the three artists he was certainly closest to 'the people'. His vigorous and personable, if rather folksy, sketches of soldiers in the 36th (Ulster) Division were effectively uniformed versions of the tinkers and shipyard workers for which he subsequently became known. In so far as Ulster aspired to the ideal of 'a nation in arms', this was it. His pictures, which included a commissioned painting of a female munitions worker at Mackies (the Belfast engineering works) and one of a nurse supporting a wounded soldier for an Ulster Volunteer Force Hospital fund raising brochure, mostly dealt with the home front or soldiers training before they went overseas, although in 1919 he exhibited a painting entitled 'The Glorious Dead'.[43]

What was seen of this war art in Ireland? The answer is not much. Conor exhibited several war-related works at the Belfast Art Society between 1914 and 1919, several of which were bought by the Belfast Municipal Art Gallery, but none of Orpen's war paintings ever came to Ireland. In May 1918 he had a major show of 125 works at Agnew's in London. The Board of the National Gallery of Ireland was 'very anxious' to have it. 'Irishmen', wrote the gallery director, 'are proud of Sir William Orpen and his pictures would help to lift some of them out of their parochialism.'[44] But in the end it was 'judged to be ill-advised to send a war exhibition to Dublin at the time', and the works went to America instead.[45] Indeed, the one main gap in the National Gallery's Orpen collection remains his war art.

Lavery gave in 1929 a very substantial collection of paintings to the Belfast Museum and Art Gallery (now the Ulster Museum) among which is a large picture entitled 'Daylight raid from my studio window, 7 July 1917'. It shows an aerial bombing raid on London[46] and we see the artist's wife looking on from within the studio. This is the only significant Great War painting in a public collection in Ireland and it provides a strikingly civilian and detached image of the war, which is reduced to a distant pattern of aircraft in conflict. It is the visual equivalent of Yeats's poem.

II

Having looked more-or-less exclusively at aspects of 'high' culture, I now want to turn to an area where art – or at least some aspects of art, especially sculpture – interacts with the local community. In the various kinds of war memorial erected following the Great War, the very specific comemmoration of the conflict, particularly by veterans and their friends and families, intersects (sometimes quite sharply) with wider public perceptions of what the war was about. To borrow Samuel Hynes's phrase, they provide a quite particular 'imagination' of the war. Here, too, are some answers to the question of *how* the war was (and is) remembered.

One of the immediate problems facing anyone who wishes to erect a war memorial is to decide whether the memorial should have any practical application or simply be a purely symbolic feature.[47] The history of the Irish National War Memorial illustrates this point.[48] In July 1919 at a meeting in the Viceregal Lodge summoned by Lord French, it was agreed to erect in Dublin a Great War Memorial Home, for current and ex-servicemen. The scheme fell into difficulties almost immediately. The military authorities objected to an institution for both serving and ex-servicemen on the grounds that discipline among the former would be difficult to maintain. The Anglo-Irish conflict added to the Trustees' problems and the Memorial Home scheme was dropped, and several others suggested in 1920–2. It was, however, agreed to proceed with *Ireland's Memorial Records* (published in 1923), an eight-volume set which aimed to list all of Ireland's war dead, with a brief biographical entry for each person. By this stage about £40,000 had been raised and the money lay rather heavily on the Trustees' hands. A few thousand pounds were spent on battlefield memorial crosses in France, Flanders and Salonika, but the chief intention remained to put up some memorial in Dublin.

In March 1924 the idea of purchasing the (private) gardens in Merrion Square, erecting some specific memorial there and then turning the resulting park over the people of Dublin was launched. Leave was obtained from the High Court to alter the original terms of the War Memorial Trust, and in 1927 private legislation was introduced in the Oireachtas (parliament). The bill passed the Seanad, through the casting vote of the Cathaoirleach (chairman), Lord Glenavy, but was withdrawn from the Dáil when it became clear that the government opposed the measure. Kevin O'Higgins, the Vice-President of the Executive Council, told the Dáil that it would be inappropriate to place such a memorial directly opposite the seat of the Irish government. It would, he declared, 'give a wrong twist, as it were, a wrong suggestion, to the origins of this State'. He argued that a visitor, 'not particularly versed in the history of the country' might come to Merrion Square and 'conclude that the origins of this State were connected with that park and the memorial in that park, were connected with the lives that were lost in the Great War ... That is not the position.'[49] There was also the problem of Irish popular perceptions. In April 1926 William Cosgrove had explained the matter to the Irish High Commissioner in London. 'A large section of nationalist opinion', he wrote, 'regards the scheme as part of a political movement of an imperialist nature and view it with the same resentment as they view the exploitation of Poppy Day in Dublin by the most hostile elements of the old Unionist class.'[50]

On 11 July 1928 the Executive Council (Cabinet) turned down a further suggestion from the Trustees of the Irish National War Memorial for a monumental arch at the main gate of the Phoenix Park in Dublin. In March the following year when the Executive Council considered a wide range of possible memorials their opinion clearly fell towards some practical scheme. The first of all their preferences was an apprenticeship scheme, which was followed by a children's educational fund and a veterans' home 'with grounds'. A memorial park came eighth out of twelve and a monument at the very bottom of the list.[51] Yet by the end of 1929 the government had agreed in principle to a memorial park. A site was provided at Islandbridge – at what the *Sunday Times* called a 'distant backwater'[52] – and Sir Edwin Lutyens engaged to design the memorial, which eventually consisted of a Cross of Sacrifice, a Stone of Remembrance, four pavilions intended to contain copies of the *Memorial Records*, pergolas, fountains and a twenty-five acre landscaped park.[53] Construction began at the end of 1931 and the scheme was completed in 1938.[54] In December Éamon de Valéra, the Taoiseach (Prime Minister), agreed to attend the formal opening ceremony which was planned for 30 July the following year. De Valéra felt that 'a ceremonial opening such as was contemplated would have a good effect by signifying that Irishmen who took different views in regard to the war of 1914–18 appreciated and respected each other's views'. But in the spring of 1939, due to the 'tenseness of the international situation' and the possibility that conscription might be applied in Northern Ireland, the moment seemed less propitious and the ceremonial opening was postponed *sine die*.[55]

The fact that the Irish National War Memorial, having begun with the intention of establishing a practical memorial, eventually opted for a symbolic scheme reflects not only the specific difficulties of setting up their originally-conceived hostel, but also a more general perception of how best to commemorate Irish war dead. In other parts of Ireland practical schemes were promoted, but in a number of instances were also supplemented with purely symbolic monuments. Technical colleges or 'institutes' of various sorts were quite popular; they were suggested in a number of places[56] and built in Limavady (Co. Londonderry) and Newry (Co. Down). In both of these places, however, monuments were later erected: a celtic cross in Limavady and a cenotaph in Newry. War memorial halls were erected in Coagh (Co. Tyrone) and Waringstown (Co. Down), and in Cavan a new operating theatre for the County Infirmary was built.[57] Quite a substantial 'Seamen's Institute' was built on Eden Quay, Dublin, as 'A tribute to the war service of seamen 1914–1919'. There was a difference of opinion in Lurgan (Co. Armagh). Early in 1919 the possibility of a technical school was being canvassed, but in the spring the Lurgan Winders' and Weavers' Union proposed a public swimming baths. In mid-April it was announced that the memorial would be a monument in the Mall. But there matters seem to have halted for the next time the project surfaced was in March 1921 when, owing to a lack of agreement on the design, the whole project had been abandoned and the subscriptions returned.[58] The matter was subsequently revived and a modest 'temple', surmounted by a bronze figure representing 'the spirit of Victorious Peace'[59] was unveiled in May 1928.

Embarrassments of this sort attended other schemes. In Cookstown (Co. Tyrone) a design for a statue had actually been commissioned, but a strong lobby held out for the construction of a hospital. In the event only just over a thousand pounds had been raised by the autumn of 1920, not enough for either proposal and the subscriptions were returned. Five years later the question was reopened and a relatively cheap (£800) and crude replica of the Cenotaph in Whitehall was eventually put up.[60] Money was a difficulty in other places. In June 1925 the 'memorial stone' was laid for the West Front of Belfast Cathedral, which was intended to be a 'Thankoffering for Victory and a Memorial to the Fallen'. Although an appeal for funds had been launched in 1919, by this stage only some £10,000 had been raised. The object of a war memorial, complained the historian of the Cathedral, 'deserved a larger response, but the citizens of the Province were distracted by rebellion and outrage and men's minds were more occupied with the difficulties and dangers of the present than the victories of the past.'[61] In November 1918 an appeal for £25,000 to erect a County Antrim memorial was launched. Eighteen years later when the memorial was taken over by the County Council only just over £3,000 had been subscribed and the work – on a reduced scale than originally planned – had only been made possible by generous donations (including a legacy in his will) from the High Sheriff of the County, Henry Barton.[62]

This sort of haphazard progress was quite typical of Irish war memorials. Their planning, commissioning and building straggled on into the 1930s, with Newry in 1939 being the last to be unveiled. At the other end of the scale the quickest to be erected seems to have been the memorial in Bray (Co. Wicklow) which was unveiled in March 1920, Matters here were apparently pushed ahead by Lord Powerscourt, chairman of the local war memorial committee.[63]

A major consideration with a war memorial is that of design. The Bray monument is a celtic cross, which was the most popular design in what became the Irish Free State, and can be found at Sligo, Nenagh (Co. Tipperary), Castlebellingham, Drogheda (Co. Louth, both) and Longford, among others. This design is much less common in Northern Ireland, but it is difficult to draw any conclusion from this except to note the Northern Protestants' traditional unease with the cross as a religious symbol. What we can say is that, with one exception, the memorials outside Northern Ireland are either crosses or abstract monuments, or a mixture of the two, as at Islandbridge.

There is a curious celtic Romanesque structure in Port Laoise (Co. Laois), an obelisk in Tullamore (Co. Offaly) and another in Cork, which has a bas-relief infantryman on one face – the only figurative memorial in the South. Soldiers in various stances from aggression to mourning are quite common in the North. But I do not think it is possible to generalise as to why the soldier in Banbridge (Co. Down) is raising his helmet and cheering (we know why he is doing this, but why in *Banbridge*?), the one in Bushmills (Co. Antrim) fiercely at the ready with bayonet fixed, or the one in Enniskillen (Co. Fermanagh) in mourning with rifle reversed.

More rewarding, however, is a study of inscriptions. What do the memorials say about why all these Irishmen died? In Ballywalter, Comber and Newtownards (Co. Down, all) they died for 'King and Country'; it was just 'Country' in Bray, Bangor (Co. Down) and Bushmills; reflecting, perhaps, the benefits of a classical education, 'Pro Deo et Patria' in Belfast. In Castlebellingham they died for 'Ireland'; and in Cork, magnificently, it was 'for the freedom of small nations'. There are only a few ideals: in Ballycastle (Co. Antrim) it is 'freedom and justice'; in Downpatrick they died for 'others'; and in Lisburn (Co. Antrim) and Portadown (Co. Armagh) they died 'that we might live'. The most common inscriptions, however, omit any specific object; most of the men appear not to have died *for* anything; that is to say, they simply 'died' or 'laid down their lives' or 'made the supreme sacrifice'.

This vagueness about purpose reflects the widespread uncertainty, in Great Britain as well as Ireland, about what the war had actually been for. The great aims of freedom, justice, liberty, democracy, and so on, did not always seem to fit easily with the observable post-war circumstances. The problem was accentuated in Ireland where the post-war conflict and partition undermined the political integrity of the war as it was remembered and commemorated. In the immediate post-war period there were attempts (as had characterised recruiting appeals during the war)[64] to link the war with the Irish national cause. On 29 July 1919 Joe Devlin, addressing a 16th (Irish) Division reunion fête in Celtic Park, Belfast, claimed that the war had been fought 'in defence of liberty for the world. Unfortunately', he continued, 'the close of the war brought to Ireland no peace and freedom, but strife and repression.'[65] The Irish Nationalist Veterans' Association, as Lieutenant J.J. Burns told the Longford Comrades of the Great War on 27 July 1919, celebrated their military service 'for Ireland and the cause of Ireland', and his association had as a specific object 'to ensure self-determination for Ireland'.[66] Such a close identification of service in the British forces and the cause of Irish nationalism, however, could not easily survive partition. Nor did it.

Unveiling a memorial in Virginia (Co. Cavan) in August 1923, Major-General Sir Oliver Nugent, a local grandee, demonstrated a striking sensitivity to the changed political circumstances. 'The day', he said, 'is not, I hope, far distant when the memory of all those of our country who gave their lives for civilisation *as we interpret it* and in obedience to *what they believed to be* their duty will be honoured and perpetuated in every town and village in Ireland' (my emphasis added).[67]

In some cases the details of war memorial inscriptions caused official concern. That on the Irish National War Memorial is notably reticent. On the Stone of Remembrance is the inscription suggested by Rudyard Kipling (who lost his only son with the Irish Guards), 'Their name liveth for evermore'; otherwise the memorial merely states (in English and, uniquely, Irish) 'To the memory of the 49,400 Irishmen who gave their lives in the Great War 1914–18'.[68] The Tom Kettle Memorial Committee quickly raised money and commissioned a (very fine) bust by Albert Power. It was completed by 1921 and the committee planned to erect it in St Stephen's Green, Dublin. But matters were held up by the disturbed political situation and a lengthy strike in the Stradbally quarry supplying the stone for the pedestal. Finally a date for unveiling was set in March 1927, but then the Commissioners of Public Works intervened and refused to allow the words 'Killed in France' to be included in the inscription, or even the last three lines from Kettle's by now famous sonnet: 'Died not for flag, nor King, nor Emperor/But for a dream born in a herdsman's shed/And for the secret scripture of the poor.' In the end the Commissioners withdrew their objection to the quotation (which, as Kettle's brother observed, could after all be interpreted as 'a general objection to Christianity'), but 'Killed in France'; was replaced by 'Killed at Guinchy 9 September 1916', with no indication of where Guinchy was, or what Kettle was doing there.[69]

In Southern Ireland the problem of the symbolism of the war memorials was exacerbated by the ceremonial surrounding them, especially in the annual November Remembrance services. In September 1926 General Sir William Hickie (President of the British Legion, Southern Ireland Area) complained about those people who were trying to turn 11 November into 12 July.[70] The actual organisation of veterans also reflected the particular Irish problem of reconciling service in the war with the new political dispensation. The Irish Nationalist Veterans' Association, although quite active in the immediate aftermath of the war, did not survive long into the 1920s. The much larger Comrades of the Great War, which became the Legion of Irish Ex-Servicemen, tried to distance itself from the British Legion, but eventually affiliated fully in the mid-1920s.[71] In Cork for some years there were two Remembrance Sunday parades, organised respectively by the Legion and the Cork Independent Ex-Servicemen's Club. Before 1925 they paraded separately to the Boer War Memorial at Gilabbey Rock. On St Patrick's Day 1925, however, a Great War Memorial, erected by the Independent Ex-Servicemen's Club, was unveiled, in a ceremony celebrated by both groups jointly. It is noteworthy, nevertheless, that the memorial was shrouded in a Union Flag before the unveiling (as was the case with the new Longford memorial on 27 August 1925)[72]. In November the service at the Cork memorial was unified and led by the Fair Lane Fife and Drum (Parnell Guards) Band.[73]

Regarding the problem of 'imperial' British demonstrations on Remembrance Day, wiser spirits in the British Legion (such as Hickie) and the Trustees of the Irish National War Memorial when they were considering the question of an official opening in the late 1930s, were very well aware of the problem and were anxious to keep everything calm and dignified.[74] By and large they succeeded, though every November in Dublin there were rowdy scenes and demonstrations raised by republicans, which was one argument for banishing the war memorial to Islandbridge. Some republicans, as Frank Ryan observed in 1934, appreciated the need to draw a clear distinction 'between ex-Servicemen commemorating their dead comrades and the Imperialist faction which exploits the dead',[75] a distinction, however, which was not always that easy to make.

In the North the commemoration of the war became intensely politicised quite quickly. The war, and the Battle of the Somme especially, were presented as a Unionist blood sacrifice: the Union permanently sealed with blood. At the unveiling of Coleraine (Co. Londonderry) war memorial on 11 November 1922, Sir James Craig bluntly told the crowd that 'those who passed away have left behind a great message to all of them to stand firm, and to give away none of Ulster's soil.'[76] War memorials in the North become quite potent symbols of loyalty to the British link and also of Ulster's *Protestant* heritage. For example, the memorial 'to the men of Ulster who fell in the Great War' – the 'all-Ulster' memorial – comprises the West Portals of St Anne's (Church of Ireland) Cathedral in Belfast and no Roman Catholic clergy assisted at the dedication in June 1927.[77] Some district memorials – for example Kilkeel, Magheralin and Donacloney (Co. Down, all) – stand in the grounds of Protestant churches, both Presbyterian and Church of Ireland. The unveiling of the City of Belfast memorial on 11 November 1929 was a similarly Protestant event (though the Belfast Rabbi participated), and was accompanied by an interesting omission from the official list of wreaths drawn up in the order of service. The 36th (Ulster) Division veterans are there (as are the Italian Fascist Party and the Ulster Women's Units of the British Fascists), but not those of the 16th (Irish) Division in which men of Catholic West Belfast tended to serve. In fact the 16th Division Ex-Servicemen's Association did lay a wreath, but virtually at the end of the ceremonies. At best this was a sin of omission on the part of the organisers, since the following year the 16th Ex-Servicemen were quite well up the 'official' list, joined that year by the Italian Fascists, the British Fascists Northern Ireland Committee, the Ulster Women's Units of the British Fascists, and the British Fascists Children's Club – quite a family outing.[78]

Not all war memorial ceremonies were by any means exclusively Protestant. For example, at the unveiling of Portadown War Memorial in November 1925, clergy of all denominations, and Sir James Craig, joined in. It was planned that the 'Dedicatory Prayer' at the start would be given by 'The Primate of All Ireland, The Most Reverend Charles F.M. Darcy, D.D.' (the Church of Ireland – Protestant – Archbishop of Armagh) and that the benediction at the end (just before the British National Anthem) would be left to 'The Archbishop of Armagh, Most Reverend Dr O'Donnell' (the Roman Catholic archbishop). In the event Dr O'Donnell did not attend and sent Canon McDonald, parish priest for Portadown, as his representative.[79] Nevertheless, the treatment of one representative group of Belfast's Catholic ex-servicemen (among other examples) illustrates the degree to which Catholic (and no doubt predominantly nationalist) ex-servicemen were sidelined and marginalised. Like Sir William Orpen they, too, became 'observers'.

To these observers we can add another group: the women. It was a man's war, yet remembered by both men and women. Among the poignant scenes each November were clusters of widows wearing their husbands' war medals. Women, too, played a significant role in raising money both for ex-servicemen's charities and the erection of memorials. They were, for example, well represented on the Dublin and District Committee of the British Legion Relief Fund,[80] and in Sligo in 1927 all twenty Legion poppy-sellers were women.[81] Their role in the commemoration, however, was not often formally noted. Clogher (Co. Armagh) war memorial is unusual in stating that it had been 'erected by fellow countrymen and women'. But the dead are male: in Ballycastle the memorial is dedicated to 'the brave men who gave their lives'; the National War Memorial is to the '49,400 Irishmen' who died; Belfast remembers 'her heroic sons'. There is an instructive contrast between the two world wars in this respect. In 1922 Lisburn war memorial was erected 'to the glorious memory of the Lisburn men who gave their lives'. After 1945 the inscription was changed: 'To the glorious memory of the men and women of Lisburn who gave their lives in two wars'.

Yet there are females at some of these memorials: peace, victory, sacrifice and grief, apparently all female characteristics, are represented by symbolic female figures. At Bangor and Lurgan 'Victory' holds up a palm frond; at Lisburn and Derry she holds both a sword and a laurel wreath. In the latter memorial, moreover, she stands with one foot upon the head of a serpent entwined around a globe. The memorial at Queen's University, Belfast, comprises a bronze group representing 'Sacrifice', in which a (female) Winged Victory sustains a stricken youth.[82] At Coleraine, perhaps indicating their relative status, on the plinth below the bronze soldier is a caped figure 'representing a daughter of Erin placing a wreath in loving remembrance'.[83]

III

The way in which the Great War was – and is – remembered and commemorated in Ireland exemplifies the often equivocal response, north and south, to the issues of patriotism, national sacrifice and personal loss which were raised during 1914–18 and in the following two decades. The comparatively limited cultural legacy, in terms of music, visual art and literature, clearly reflects the lack of any coherent or whole-hearted Irish commitment to the war, or, indeed, *against* it. In a sense, the war did not matter to Ireland. Despite efforts to equate Ireland with Belgium, and John Redmond's attempts to fire *Irish* patriotism in support of the Allied war effort,[84] during the war itself and after there was a collective lack of engagement with the conflict. Although a great number of Irishmen volunteered to fight, and very many died, Ireland as a whole – or at least nationalist Ireland – was detached from the war, remaining, like Orpen and Lavery, and Yeats's airman, in a sense 'onlookers'. After 1918 this distancing from the war became inevitably more marked. It is nowhere more aptly illustrated than in the progressive re-location of the proposed Irish National War Memorial from a central Dublin hostel in 1919, to a cenotaph in Merrion Square five years later, to the gates of the Phoenix Park in 1928, and finally to the 'distant backwater' of Islandbridge the following year. The memorial was, literally, removed from the centre of attention.

The history of Great War memorials and Remembrance ceremonies in Ireland generally during the two decades after 1918 further illustrates a considerable degree of collective disengagement from the commemoration of the war. Republicans, of course, strongly opposed what they saw as an annual affirmation of British imperialist values in the November ceremonies. The inter-war administrations of the Irish Free State – both Cumann na nGaedheal and Fianna Fáil – were less categorical. Although the government refused to permit the erection of the National War Memorial in central Dublin, it did provide the eventual site at Islandbridge and also contributed substantially to the cost of building the memorial park. While de Valéra's agreement to attend the dedication of the National War Memorial was overtaken by events, a government representative attended some of the inter-war annual Armistice Commemorations in Dublin. From 1923 to 1936, moreover, the Irish government was officially represented every November at the Cenotaph service in London.[85] In Northern Ireland, however, there was no official hesitation at all in commemorating the war, since in the 1920s particularly, when most war memorials were being dedicated, it provided repeated opportunities to affirm Britishness and imperial loyalty. For veterans and the war-bereaved, however, matters were less clear-cut. For a number of reasons Catholic – no doubt predominantly nationalist – people in the North were not always fully represented at these occasions. Many in the South, too, were uncomfortable with the imperial symbolism which accompanied Remembrance services. For these groups of people the easiest option was perhaps simply to turn away from the formal and public commemoration of the war and its sacrifice. We cannot tell what the personal cost of such a response might have been, but the public cost was substantial. This was especially so in Northern Ireland where an appreciation of the complex and subtle range of meanings which could be drawn from the common tragic experience of the war was overwhelmed by a simple patriotic and predominantly Protestant type of commemoration. Although some people made commendable attempts to stress the common sacrifices made by virtually all-Ireland during the war,[86] the sombre truth remains that the nationalist and unionist Irish casualties of the Great War became more divided in death than they had ever been in life.[87]

NOTES

1   Oxford, 1975.

2   London, 1990.

3   Dublin, 1986. David Fitzpatrick has himself stimulatingly addressed the subject in, for example, *Politics and Irish Life, 1913–21* (Dublin 1977), and 'The overflow of the Deluge: Anglo-Irish relationships, 1914–1922', in Oliver MacDonagh & W.F. Mandle (eds), *Ireland and Irish-Australia: Studies in Cultural and Political History* (London 1986), pp. 81–94.

4   Terence Brown, *Ireland: a Social and Cultural History, 1922–79* (Fontana edn, London 1981), p. 123.

5   Biographical details from the *Dictionary of National Biography*; entry on Stanford in *New Grove Dictionary of Music and Musicians* (London 1980); and Frederick Hudson, 'A Revised and Extended Catalogue of the Works of Charles Villiers Stanford (1852–1924)', *Music*

Review, xxxvii (1976), pp. 106–29 (I am grateful to Dr Desmond Hunter of the University of Ulster for drawing my attention to this article).

6   Quoted in Hynes, *A War Imagined*, p. 17.

7   Malcolm Ruthven, *Belfast Philharmonic Society, 1874–1974* (Belfast 1973 [*sic*]), p. 61.

8   Biographical details from David Greer (ed.), *Hamilton Harty: His Life and Music* (Belfast 1979).

9   The complete poem is printed in the Novello edition of the cantata (London 1913).

10   *Belfast News-Letter*, 27 Nov. 1919. It is perhaps also worth noting that there seems to have been no current prejudice against German music and that this first concert after the end of the war also included 'a very enjoyable rendering of the overture to "The Meistersingers" (Wagner)'.

11   Biographical details from *New Grove*.

12   Stainer & Bell edition (London 1920).

13   Entry on Stanford in *Dictionary of National Biography*.

14   Modris Eksteins, *Rites of Spring: the Great War and the Birth of the Modern Age* (London 1989).

15   John F. Larchet, 'Music in the Universities', in Aloys Fleischmann (ed.), *Music in Ireland* (Cork & Oxford 1952), p. 18.

16   14 Feb. 1930, 30 Nov. 1945 and 7 Nov. 1958. Ruthven, *Belfast Philharmonic*, pp. 62–3.

17   See *Oxford Dictionary of Quotations* (2nd edn revised, Oxford 1966), p. 571:1.

18   The question of Irish racial stereotypes in World War I is discussed in Terence Denman, 'The Catholic Irish soldier and the "racial environment" in World War I', in *Irish Historical Studies*, xxvii, no. 108 (Nov. 1991), pp. 352–65.

19   The title of the *'Daily Express' Community Song Book No. 3* (London 1930), collected and edited by S. Louis Giraud.

20   Grand Opera House programme, Monday 3 Mar. 1930 (Public Record Office of Northern Ireland [henceforward PRONI] T.1933).

21   Hynes, *A War Imagined*, p. 442.

22   J.C. Trewin, 'Introduction' to *Three More Plays by Sean O'Casey* (London 1965), pp. x–xi.

23   Sean O'Casey, *Rose and Crown. Autobiography: Book 5* (Pan edn, London 1973), pp. 103–4.

24   Trewin, 'Introduction', p. xii.

25   Patrick F. Sheeran, *The Novels of Liam O'Flaherty* (Dublin 1976), p. 67.

26   Hugh Cecil, 'The Literary Legacy of the War: the Post-War British War Novel – a Select Bibliography', in Peter Liddle (ed.), *Home Fires and Foreign Fields: British Social and Military Experience in the First World War* (London 1985), p. 220.

27   Sheeran, *Novels of Liam O'Flaherty*, p. 67.

28   *Return of the Brute* (London 1929), p. 137. At about the same time O'Flaherty also wrote 'The alien skull', a similarly grimly realistic short story set on the Western front, first published in *The Mountain Tavern* (London 1929), pp. 163–74.

29   'An Irish Airman Foresees His Death', in *The Collected Poems of W.B. Yeats* (2nd edn, London 1950), p. 152.

30   Including, for example, any work by Wilfred Owen. Quotations from Yeats's introduction to *The Oxford Book of Modern Verse, 1892–1935* (Oxford 1936), pp. xxxiv – xxxv.

31   George Bernard Shaw, *Selected One Act Plays* (Harmondsworth 1972), p. 68.

32   Sean Keating, 'William Orpen: a Tribute', *Ireland Today* (1937), quoted in Bruce Arnold, *Orpen: Mirror to an Age* (London 1981), p. 301.

33   *An Onlooker in France, 1917–1919* (London 1921; revised and enlarged edn 1924).

34   Ibid. (1924 edn), p. v.

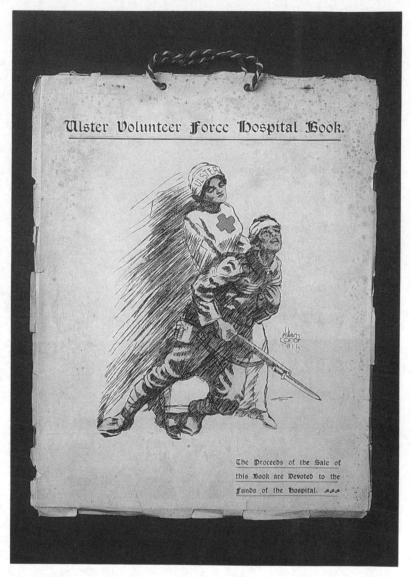

William Conor, from a fund-raising booklet in aid of the Ulster Volunteer Force Hospital
(Linen Hall Library, Belfast)

William Orpen, 'To the Unknown British Soldier in France' (original version)
(Imperial War Museum, London)

William Orpen, 'To the Unknown British Soldier in France' (final version)
(Imperial War Museum, London)

35   Arnold, *Orpen*, p. 359; Meirion & Susie Harries, *The War Artists* (London 1983), pp. 146–9.

36   Interview with Sir William Orpen, *Evening Standard*, 7 May 1923 (Imperial War Museum [henceforward IWM], World War I war artists press cuttings book).

37   Quoted in Arnold, *Orpen*, pp. 379–80.

38   Charles ffoulkes (Curator & Secretary, IWM) to Sir Martin Conway (Director-General, IWM), 3 Jan. 1923 (IWM Dept. of Art 1st World War archive: Sir William Orpen [henceforward IWM: Orpen] 78C–3, fol. 90).

39   *Daily Herald*, 8 May 1923; *Patriot*, 10 May 1923 (IWM World War I war artists press cuttings book).

40   See correspondence with Orpen, Feb.-Apr. 1928 (IWM: Orpen, 78D–3, fol. 42, 46, 56, 60–1).

41   Harries, *War Artists*, p. 132; Kenneth McConkey, *Sir John Lavery R.A. 1856–1941* (Exhibition catalogue, Belfast & London 1984), p. 76.

42   Harries, *War Artists*, p. 36.

43   Judith Wilson, *Conor 1881–1968: the Life and Work of an Ulster Artist* (Belfast 1981) contains details of Conor's World War I work.

44   R. Langton Douglas to Alfred Yockney (Ministry of Information), 25 June 1918 (IWM Dept. of Art archives, file 47/2).

45   Arnold, *Orpen*, pp. 340–1.

46   McConkey, *Lavery*, p. 83 describes the picture as commemorating 'the second aerial bombing raid on London and showing an encounter between aircraft and airships'. In fact it was the second (and last) daylight raid by fixed-wing German aircraft (there had been earlier attacks by Zeppelins), and no airships appear to have been involved. See A. Rawlinson, *The Defence of London, 1915–1918* (London 1923), pp. 177–8.

47   There is a growing literature on war memorials. See Alex Bruce, 'War memorials', in *The Historian*, no. 28 (Autumn 1990), pp. 17–20, for an introductory survey.

48   There is an excellent summary of the history of this memorial in Jane Leonard, '"Lest We forget"', in Fitzpatrick, *Ireland and the First World War* (Mullingar 1988), pp. 59–67. The early history of the memorial was reviewed in a Seanad debate on 9 Mar. 1927 by Andrew Jameson, a trustee of the Irish National War Memorial (*Seanad Éireann Official Report*, viii, col. 421–32).

49   *Dáil Éireann Official Report*, xix, col. 400 (29 Mar. 1927).

50   Cosgrove (President of Executive Council) to James MacNeill, 8 Apr. 1926 (National Archives of Ireland [henceforward NAI], D.T. [Department of the Taoiseach papers] S4156A).

51   Note from President to Cabinet ministers, 2 Mar. 1929 (ibid).

52   *Sunday Times* (London), 31 Aug. 1930 (clipping filed in NAI, D.T. S4156B).

53   The Cross of Sacrifice, designed by Sir Reginald Bloomfield, and Lutyens' Stone of Remembrance were common to all the principal British Great War cemeteries and some other memorials in France and Belgium. See T.A. Edwin Gibson & G. Kinsley Ward, *Courage Remembered* (London 1989), pp. 52–4.

54   See *British Legion Annual 1941*, special number on the Irish National War Memorial (copy in Irish Architectural Archive, Dublin).

55   Note regarding opening of Irish National War Memorial, 28 April 1939 (NAI D.T. S4156B).

56   Including Bangor, Lurgan and Portadown. See *Irish Builder*, lx, no. 22 (30 Nov. 1918); lxi, nos. 1 & 3 (11 Jan. & 8 Feb. 1919).

57   Ibid., lxii, no. 26 (4 Dec. 1920).

58   Ibid., lxi, nos. 1, 7 & 8 (11 Jan., 5 & 19 Apr. 1919); lxiii, no. 6 (12 Mar. 1921).

59   As described in the 'Order of Service for Unveiling and Dedication, 23 May 1928' (copy in IWM National Inventory of War memorials).

60   See Cookstown War Memorial Committee Minute Book (May 1919-Apr. 1927) (PRONI LA 28/16AB/1).

61   Judge Thompson & F.J. Bigger, *The Cathedral Church of Belfast* (privately published, Belfast 1925), p. 17.

62   Doreen Corcoran, 'County Antrim War Memorial at the Knockagh', *Carrickfergus & District Historical Journal*, ii (1986), pp. 17–19.

63   Leonard, '"Lest We Forget"', p. 62.

64   See Mark Tierney, Paul Bowen & David Fitzpatrick, 'Recruiting Posters', in Fitzpatrick, *Ireland and the First World War*, pp. 47–58.

65   *Belfast News-Letter*, 30 Jul. 1919.

66   *Longford Leader*, 2 Aug. 1919.

67   Nugent Papers (PRONI D. 3835/E/7/23).

68   This inscription was specifically approved by the Cabinet. See Extract from Cabinet Minutes, 19 Feb. 1935 (NAI D.T. S4156B). The figure of 49,400 is a considerable overestimate. Kevin Myers had suggested that a figure in the region of 35,000 would be more appropriate. See Patrick Callan, 'Recruiting for the British Army in Ireland During the First World War', *Irish Sword*, xvii, no. 66 (Summer 1987), pp. 42–56.

69   J.B. Lyons, *The Enigma of Tom Kettle* (Dublin 1983), pp. 305–6.

70   Hickie was addressing a British Legion meeting in Athy on 22 Sept., *Irish Times*, 23 Sept. 1926. Kevin O'Higgins (Vice-President & Minister of Justice) made a similar point in the Dáil debate on the Merrion Square memorial scheme, 29 Mar. 1927, *Dáil Éireann Official report*, xix, col. 403.

71   There was also a 'Roger Casement Brigade (Survivors)' association which existed in the mid-1930s. See correspondence in NAI D.T. S8817.

72   *Longford Leader*, 29 Aug. 1925. Photographs of the unveiling were reproduced in '50 Years of Longford 1936–86', *Longford News* special number, Dec. 1986.

73   For the Cork Remembrance services and memorial, see *Cork Weekly Examiner*, 15 Nov. 1924, 21 Mar. & 21 Nov. 1925.

74   See, for example, notes of meetings with representatives of ex-servicemen, 25 Aug. 1938 & 27 Apr. 1939 (NAI D.T. S4156C).

75   Quoted in Sean Cronin, *Frank Ryan: the Search for the Republic* (Dublin 1980), p. 60.

76   *Belfast Telegraph*, 11 Nov. 1922.

77   *Cathedral of Saint Anne, Belfast: the Form and Order of the Dedication of the Portals of the West Front*, 2 June 1927 (copy in Belfast Central Library, Irish Collection). The completion of the West Front as a war memorial was apparently Lord Carson's idea. See C.E.B. Brett, *Buildings of Belfast* (London 1967), p. 57.

78   See *The Unveiling of the Belfast War Memorial*, 11 Nov. 1929 (brochure in Ulster Museum Local History Collection), and *Belfast Weekly Telegraph*, 16 Nov. 1929 & 15 Nov. 1930.

79   Wreaths were laid by representatives of both the local Orange Order and the Ancient Order of Hibernians. *Portadown and District War Memorial: Unveiling and Dedication*, 13 Nov. 1925 (copy in IWM Department of Printed Books); *Belfast Telegraph*, 14 Nov. 1925.

80   Seven out of the seventeen members at a committee meeting in 1926 were women. *Irish Times*, 13 Oct. 1926.

81   *Sligo Independent*, 12 Nov. 1927.

82   See *Irish Builder*, lxviii, no. 24 (5 Nov. 1921); Robert Marshall, *The Queen's University of Belfast Services Club, 1918–1968* (Belfast 1968), p. 15.

83   *Belfast Telegraph*, 11 Nov. 1922.

84   See, for example, his Introduction to Michael MacDonagh, *The Irish at the Front* (London 1916), pp. 1–14.

85   A representative attended in Dublin in 1924 and 1928–31 inclusive. See 'Note regarding Government representation at Armistice Day Ceremonies' (1937) (NAI D.T. S3370B).

86   The fact that, when he was fatally wounded in France in June 1917, Major William Redmond MP was brought in by stretcher-bearers from the 36th (Ulster) Division provided an opportunity for reflections of this sort. See, for example, the oration by Thomas O'Donnell ('There in that stricken camp was buried for ever the prejudice of ages.') at the eighth anniversary celebrations in memory of Redmond, Ennis, County Clare, *Cork Weekly Examiner*, 13 June 1925.

87   I am particularly grateful to the following people who have helped in the preparation of this essay: Fred Jeffery, Sally Visick, Jane Leonard, Brian O'Hara, Patrick Maume, Ken Inglis, Mary Morrissey and Peter Johnson, whose collection of historic picture postcards provided much vital evidence.

ARMY, POLITICS AND SOCIETY IN INDEPENDENT IRELAND, 1923–1945

Eunan O'Halpin

This paper discusses the domestic intelligence activities of the Irish army during two extended periods of national crisis, the early 1920s and the Emergency from 1939 to 1945. The scope and duration of military involvement in domestic, political and security issues suggest a more complicated set of relationships between the Irish army, government and people than is usually described.

One of the badges of military professionalism is a certain disdain for conventional politics, coupled in a democracy with acceptance that the army is the servant, not the master, of elected government.[1] Irish politicians have usually treated the army with a mixture of affection and calculated neglect. It has existed and developed in a vacuum, because, apart from hasty and most secret arrangements for co-operation with the British in the event of a German invasion in 1940, no government has been willing to endorse a remotely credible external defence policy. This raises the question of what politicians think the army is there for at all. Two fairly recent episodes show how sensitive a point this is.

In 1976 the 'convivial and ebullient' Minister for Defence, Patrick Donegan, in the course of a bizarre attack on the incumbent President O'Dalaigh, gave the conventional view of the army's apolitical position in the phrase: 'The army must stand behind the state'. During a further political crisis centred on the presidency in 1990, the Taoiseach, Charles Haughey, responded with uncharacteristic emotion to opposition accusations that he had threatened the promotion prospects of an army officer.[2] Yet as Mr Donegan and Mr Haughey surely knew, the army has at times been deeply embroiled in domestic political affairs, while the careers of senior army officers have notoriously frequently been influenced by political considerations. The problem to be considered is why there is such a gulf between the formal and the practical use made of the Irish army by governments since independence.

The overall explanation is plain enough: the greatest threat to the Irish state since independence has been an internal one, whereas the conventions of democratic politics and of military professionalism suggest that armies exist to defend countries from external attack, to project national power externally, and to contribute in some way to the maintenance of international order. In terms of any of these criteria the Irish army has never been much more than a simulacrum. However, it has been highly effective in defending the state from within. Simply by its existence the army has been a bulwark against a republican seizure of power. At times of crisis, furthermore, it has been very active in internal security operations, sometimes in ways on which neither army nor government wish to dwell. Such work is inherently sensitive. It invites the charge that the army is being used for domestic political purposes, to deal with opponents of the government rather than enemies of the state. It suggests that the army in practice is an internal security force, and thus may diminish its standing as a focus for national loyalty. It may also bring to the surface profound political divisions within what is notionally an antiseptically non-political professional body. This happened in the first two decades of the state's existence: senior army officers were involved in a half-baked mutiny in 1924, a vague plan for a *coup d'état* in 1931–2, talk in the mid-1930s of launching a war against Britain to reclaim Northern Ireland, and in intrigues with a fugitive German agent in 1940–1. These follies came to light and were dealt with partly through the efforts of fellow officers with a stronger sense of constitutionality and a firmer grasp of strategic reality. Such episodes suggest that, at least for the state's first twenty years, no government could be absolutely certain of the complete loyalty of the defence forces. Keeping the army out of politics, and politics out of the army, was never so simple a matter as is sometimes assumed. This consideration must be kept in mind in the discussion which follows of the army's notional role and actual functions in internal security during that time.

During the years of the Emergency from 1939 to 1945 the Irish army exercised extraordinary powers over national life. The military had a far broader remit and were far less constrained by law, by custom, by civil government or by rival institutions than the British or American wartime security agencies and armed forces.[3] This was because during the Emergency the government, in addition to the primary defence task of mobilising the nation to defend Irish neutrality from external threats, gave the army extremely broad internal security responsibilities. These were discharged with efficiency and discretion by G2, the army's military intelligence directorate.

G2's acquisition of a pivotal role during the Emergency did not happen by chance. It was nevertheless somewhat surprising. As we shall see, army intelligence had taken the lead in domestic intelligence work during the state's first years, when the republican movement, including the political wing led by Mr de Valéra, was its main target for penetration and scrutiny. Yet it was Mr de Valéra's government which gave the army such broad powers, and such discretion in using them, once external threats to the declared neutrality policy were recognised late in the 1930s.

The first unequivocal military warning on the internal security implications of neutrality appears to be that in the G2 document 'Fundamental Factors Affecting Irish Defence Policy' of May 1936. This important paper was, according to its principal author Dan Bryan, then deputy director of G2, directed primarily against the 'utter insanity' of a group of senior officers who were 'talking extensively about a military war against the British and the successful manner in which such a war could be waged'.[4] 'Fundamental Factors' helped to scotch this line of thought. It emphasised the importance of Ireland's strategic position in the Atlantic, the country's total dependence on British seapower for the protection of her trading links, the wartime implications of Britain's retention of the treaty ports, and the fact that Northern Ireland remained a part of the United Kingdom and consequently would be involved if Britain went to war. It analysed British Atlantic naval operations based on Ireland during the First World War, and stressed Germany's unqualified diversionary success in fomenting rebellion in Ireland in 1916. It pointed out that since 1921 the British had repeatedly stressed that Ireland must never allow herself to be used by an enemy to harm Britain's interests, and that extensive co-operation would be demanded from the Irish on various matters including press and communications censorship and other aspects of 'the ... larger question of Internal Security (protection against espionage, hostile propaganda, sabotage and various other undesirable activities)'. This would necessitate 'an effective national machinery'. The document also remarked that Ireland would 'always' have 'an active if small minority that can be exploited' by a foreign power aiming to make trouble for Britain.[5] Although apparently considered too politically sensitive – and at eighty pages perhaps too long – to enjoy the status of an official army paper, 'Fundamental Factors' was widely circulated as a military intelligence document. It was also used by the chief of staff in negotiations with ministers about expanding the defence forces.[6]

Those discussions proved fruitless in terms of the army's aim of securing the money, equipment and above all the lines of policy with which to begin constructing its version of a credible system of national defence. However, the increasing likelihood of a European war did spur the government into a period of intense diplomacy which culminated in the Anglo-Irish agreement of 1938. It made the maintenance of Irish neutrality in war feasible by ending British use of the 1921 treaty ports and by producing unequivocal Irish guarantees that Ireland would not be used to harm British interests in any way. As 'Fundamental Factors' had anticipated, such a commitment was going to involve elaborate co-operation with the British. A new security apparatus in Ireland would also be needed to tackle the problem of third-party use of Ireland to attack the United Kingdom, particularly where no crimes were being committed under Irish law but where Ireland's external relations were affected. Shortly after the transfer of the ports the British warned the Irish government 'that the espionage activities of Germany extended to Ireland and that these ... affected British, United States and French defence interests'. In 1945 Dan Bryan wrote that because 'similar' counter-espionage and security functions 'in almost all other States' were dealt with by defence ministries, the job was given to the army. G2 set up a 'defence security intelligence' section, to operate in collaboration with the Post Office Special Investigation Unit and a small aliens section within the Gárda Special Branch established for counter-espionage work.[7]

This was an important decision. It was not preordained that the army should be given the task. The Gárdaí had had responsibility for all internal security work since 1926. From a Fianna Fáil point of view, the Special Branch was sound – during their first years in power Mr de Valéra's government had taken some trouble to ensure the political loyalty if not the increased efficiency of its detectives, by importing into it a number of active republicans whose previous experience with the law had been in breaking it.[8] The government was also fairly confident that the army could be relied on, but Fianna Fáil ministers might justifiably have remained wary of military intelligence because of its activities in the first years of the state. These we must now discuss.

When the civil war petered out in the summer of 1923 the army had been the only institution organised on a national basis capable either of imposing order or of obtaining reliable information on political conditions and on threats to the state. The Gárda Síochána, the new civilian police, had a disastrous start. Its men were unarmed. While this contributed to its public acceptability, at a time when Ireland was awash with gunmen, some of them republicans, some demobilised soldiers, and some freebooters – such as 'a tinker from Granard named "Piper" Reilly with 3 confederates' who availed of conditions to engage in 'armed looting' – it severely limited the force's capacity to deal with disorder.[9] In the capital the hapless Dublin Metropoli-

tan Police (DMP) was in the same situation. During the civil war the government had established a number of plain-clothes armed units, some civilian
and some military, to counter various forms of republican activity. In Dublin
these included the notorious CID or 'Oriel House', the Protective Corps, the
Citizens Defence Force, and the President's Escort, and in Cork an army
'Plain Clothes Squad ... Their duties include escorting Postmen, keeping
close observation on Banks, Post Offices, Public Institutions etc.'.¹⁰ These *ad
hoc* creations bore more relation to Collins's War of Independence 'Squad'
than to detective or intelligence organisations. They had no statutory existence, they operated outside normal chains of command, they earned ugly
reputations even amongst government supporters, they were manifestly
unsuitable either for conventional police work or for gathering political intelligence, and in the aftermath of civil war many of their members were a conspicuous menace to public order themselves.

The records provide many illustrations of this. To take one example, on
30 June 1923 a Dublin taxi firm was raided and one of its cars taken by three
armed men, one of them in army uniform, who were then arrested by troops
but not charged. Two days later one of the firm's drivers was 'beaten by men
who, he alleges, were from Oriel House'. The firm's owner attempted to
reach Oriel House to complain, but 'the same men intercepted him in
O'Connell Street and fired shots over his head'. An army intelligence officer
reported that the perpetrators were 'some of the recently discharged CID
men', together with Joseph Owens, a former commander of the President's
Escort, 'a very erratic type of gentleman, as I have known him to do holds up
[*sic*] for the purpose of stealing Motor Cars, drink etc, even while he was serving in the Army ... if this gang is not broken up, the robberies will continue
indefinitely.' Two days later the army captured Owens and others in the act
of armed robbery in County Westmeath. The prisoners were taken to Dublin
to be charged, but there 'some of them were recognised as CID men' and
promptly released.¹¹

For almost two years after the end of the civil war the army was obliged
to act not so much in aid of the civil power as in lieu of it, taking action
against both the 'Irregulars' and armed criminals through raids, searches, and
arrests. In the midst of this chaos, army intelligence, which during the civil
war had been largely a flag of convenience for groups of plain-clothes soldiers
in the various army commands rather than an effective centrally controlled
organisation for the collection and analysis of information nationally, prospered. Intelligence staffs reporting to headquarters were appointed in each
command, and the headquarters organisation was strengthened. For some

months army intelligence was 'engaged solely on work in connection with Irregular activities', on which it built up an elaborate and meticulously indexed set of records including around twenty five thousand files on individual republican suspects. It developed extensive networks of paid and unpaid agents in Ireland and Britain who provided intelligence on republican activities, it collected similar information from the United States and from European countries, and at the end of 1923 it 'took control of the remnants of the Intelligence Service which existed in the North East [Northern Ireland]'.[12] So far from becoming absorbed in strictly military matters, therefore, army intelligence was largely concerned with political affairs.

The first threat to the enhanced position of the army's military intelligence branch came early in 1924, when intelligence outside headquarters was 'for all practical purposes wiped out' as part of an army reorganisation and demobilisation scheme.[13] This may connote a democratic impulse to take the army out of politics by removing its capacity for domestic spying. On the other hand, the veteran intelligence specialist Dan Bryan thought it simply the product of the new state's undue reverence for British forms and practices: he attributed the decision to 'somebody who was looking at British Peace Tables of Organisation and not conditions in Ireland'.[14] Whatever the reason, the decision did not stand for long. Intelligence made a dramatic recovery following the 'army crisis' or mutiny which the reorganisation scheme provoked amongst a faction of officers grouped around Liam Tobin, Michael Collins's right hand man during the war of independence. Because 'the [intelligence] Department was seriously affected internally it was not possible to openly deal with the matter' before the crisis erupted, but 'a very considerable amount of work' was discreetly done by loyal officers against their colleagues, who proved hopeless conspirators.[15] The outcome of the mutiny is usually seen as a turning-point in Irish political development, the moment when the army finally recognised its subordination to civil government and accepted the role of a professional military force primarily concerned with external defence. The clean out of the army which followed certainly enabled the reconstruction of its intelligence organisation on more professional military lines. However, it continued to be the government's main source of intelligence on and analysis of the republican movement, as well as on political conditions generally in the country.

Two new sections were formed within intelligence in response to the mutiny, one 'to deal with the internal supervision of the Army' and the other to watch 'elements until then in or connected with the Army, but now working for its disruption'. This justified the reappointment of intelligence staffs in the various commands, enabling the resumption of surveillance of republican organisations as well as of the disgruntled 'Tobinites'. As 'most of the [Tobinite] leaders had ... considerable experience of the Army and Governmental machine, and in particular of Intelligence methods ... the position ... resolved itself into a battle between their Intelligence service and ours (in which of course a number of them had been serving until a short time previously).'[16] By the end of the year, that struggle was over: the army had been purged of 'Tobinite' undesirables, as well as republican sympathisers and one NCO discovered to have had 'dealings with the Six County Authorities', while the various ex-servicemen's bodies no longer posed a threat. For the next year the intelligence organisation, by then known as the 'Second Bureau', enjoyed a period of expansion, despite the dramatic contraction in the size of the army. The emergence of subsections dealing with, for example, 'foreign armies', 'armaments production', and 'developments of war' suggest an intention to evolve what the director of intelligence termed a 'real' military intelligence organisation, one that would concentrate on purely military matters (See figure below).[17]

In practice, however, the Second Bureau maintained its close watch on political issues and conditions. These sometimes had an external dimension.

MILITARY INTELLIGENCE
SECOND BUREAU, 1925

DIRECTOR

DEPUTY DIRECTOR

A SECTION	B SECTION	C SECTION	D SECTION	PERSONAL STAFF
- Index of records	- Statistical	- Press Relations	- Foreign Armies	- Finance
- NE Ulster	- Field Intelligence	- Publicity	Development	Codes &
Foreign	- Topographical	Press	of War	Cyphers
Elements		Censorship	Technical	- Secretariat
- Irregulars			Information	- ex-army bodies
- Communists etc.			- Inventions	Boy Scout
- Labour			Armaments	Bodies
- Economics			Production	Organisation
			International	of service
			Matters	Interior
			Foreign	Economy
			Missions	Custody
				of orders

For example, the rooting out of the 'Tobinites' enabled increased attention to be paid to 'Communists and other extreme Labour Organisations' which had links abroad. As the Gárdaí were later also to discover, Soviet-inspired communism was a hopelessly anaemic though verbose threat to the established order.[18] However, captured documents revealed that the IRA had established communications with Moscow in 1925, and that the IRA's commanding officer in Britain had subsequently engaged in military espionage for the Soviets. This incident, isolated though it was, was of considerable significance, because it showed that a foreign power could operate against Britain indirectly through Ireland or Irish organisations.[19] Much effort was expended in assessing events in Northern Ireland: in addition to watching northern republicans, the bureau's north-east section collected 'data on all matters related to the North-East and its Government, and also ... data re hostile British elements in the Saorstat'. It watched 'the Police Force, RUC, Specials, etc.', as well as 'British troops', obtaining information on their 'strength, organisation, armament, location of units, etc. This was done largely through the use of agents.' A 'safe but secret line of communication has also been maintained with Belfast and other parts ... at all times.' The section had done more than simply amass background information for future study: it claimed to have obtained 'inside information' with which the Irish delegates on the Boundary Commission 'successfully rebutted North-Eastern evidence'.[20] The Second Bureau's most important work, however, remained its surveillance of political developments in the twenty-six counties.

This surveillance was operated by various means, most importantly through large numbers of agents who reported to army intelligence officers on all manner of political and military developments within the republican movement, as well as in Labour organisations. A selection of reports dealing with political issues from about forty of these agents – including 'CA153', '70', '140 Sligo', and 'L[iver]P[ool] 11' – together with some monthly intelligence commentaries, can be seen in the papers of Desmond Fitzgerald, Minister for External Affairs until 1927. While some of the contents are fantastic – in November 1924 a Glasgow agent wrote that 'a very reliable source' claimed 'that de Valéra has received a large sum of money from a Mexican source' – the reports chronicle in considerable detail the affairs of the government's opponents, particularly the growing tension within the republican movement on the issue of abandoning the policy of abstention from the Dáil.[21] The impression left by these documents is that amongst the army's informants were close associates of some of the leading figures in republican politics. For example, the director of intelligence passed on an account of 'a meeting of

Irregular TDs' where 'the question of entering the Dáil was again discussed' in November 1925. The report came 'from what is usually a reliable source in close touch with the opinions of Gerald Boland TD ... It is requested that very particular precaution be taken to safeguard this information in the interests of the source from which it came'.[22]

It was coincidental, but perhaps just as well for future relations between army and government, that the founding of Fianna Fáil occurred in May 1926, six weeks after the Second Bureau formally relinquished responsibility for political surveillance, and with it their well placed agents, to the Gárdaí. That decision was taken in the autumn of 1925, apparently at the insistence of Kevin O'Higgins, the Minister for Justice. Relations with the civil police, particularly with the detective unit initially established under David Neligan in 1923 and reconstituted as the Special Branch in 1925, had always been difficult. Second Bureau officers maintained that, whereas they provided relevant information freely for the Gárdaí, this was never reciprocated: one commented in November 1924 that the police 'are not even as useful to us as a friendly civilian'.[23] There were obvious dangers of overlap in the employment of informers, and there were key differences of approach. The Second Bureau disparaged the Gárdaí's propensity to raid and to arrest the small fry of the republican movement, claimed that the Gárdaí had no system for recording and collating information, and warned that 'a serious outbreak of crime' would be enough to make the police 'turn their whole attention to their normal work ... and drop or neglect Irregularism'.[24] Only the Second Bureau had the experience, the records, and the networks of informers necessary to maintain effective surveillance on organisations and tendencies which potentially were threats to the state. In September 1925 the director of intelligence asked for arrangements 'ensuring a close system of co-operation between the Civil and Military forces of the country', in place of the friction which prevailed. As things stood, the Second Bureau was frequently left in ignorance of important discoveries such as caches of republican weapons and documents, while the police also withheld 'reports made on the Organisations in the Industrial and Communistic circles which may, under certain circumstances, call for intervention by the military ... the Army is thrown back on its own resources to obtain its knowledge of these bodies'. In his view

occasions when the army may be called upon to support the Civil Authorities, or to take over their duties, cannot be regarded as distinct possibilities only, owing to the likelihood of industrial disturbances, riots or other outbreaks of violence, or to the constant menace of the militant Republican groups'.[25]

No material has come to light showing the opinions of ministers on the question, but these somewhat alarming views of the army's likely role in national affairs evidently cut no ice. The government was dominated by Kevin O'Higgins, who as Minister for Justice was almost as hostile to the army as to the republicans and who wanted to establish a conventional democratic policy. The civilians won the argument hands down. The Minister for Defence gave 'verbal instructions' that the Second Bureau hand over all its records on republican suspects, together with control of all its agents and informants, to the Gárda Special Branch.[26] Action on this took some time, apparently because of military fears that 'if the Police took over Intelligence, they would merely be concerned with the getting of guns, dumps, etc. & not ... with the obtaining of information on the *policy* of anti-State organisations'.[27] However, the transfer process finally began in February 1926. In addition to documents on individual republicans other papers were eventually handed over, including subject files dealing with 'Ulster and Anti-National elements here' such as 'Anti-Irish elements in Saorstat', 'Secret Service – British and NE Ulster', and 'Freemasonry'. These incidental transfers marked the effective abandonment of the army's claim to continued responsibility for counter-espionage and for intelligence on Northern Ireland, neither of which much interested the Gárdaí, as 'it would be very hard to get going again' without the records.[28]

Although a certain amount of political intelligence continued to dribble in – the Eastern Command monthly intelligence report for July 1926 included an account of discussions between de Valéra and republican TDs on the relative positions of Fianna Fáil and Sinn Féin on various issues – the army wound down its domestic intelligence gathering very swiftly. In April only four 'paid agents' remained, and by December the number was down to two, one of whom was based in Northern Ireland.[29] The switch in activities can be measured by the change in the Second Bureau's budget. In 1925 it was almost eleven thousand pounds, eighty per cent of which came from the secret service vote, used mainly as a fund for informers; in 1926 spending dropped to about three thousand pounds, only a quarter of which was from the secret service vote. The army's share of the overall secret service vote also dropped immediately by almost two-thirds, from 76 to 30 per cent of the total spent. While at least one army intelligence officer, Dan Bryan, remained informally in touch with some of his republican and Labour contacts, the Second Bureau obeyed its instructions to get out of internal security altogether.[30] Battalion intelligence officers were expected to keep abreast of developments in their areas, but a prohibition on the use of informants meant that whereas 'a year ago our work was almost entirely political ... nowadays we know nothing more than the ordinary man about political matters, not even receiving' police information as the Gárdaí showed no inclination to keep the military in the picture.[31]

It is clear that the decision to take the Second Bureau out of domestic intelligence gathering was correct both in principle and in practice. By 1926 the Gárdaí were sufficiently well organised to collect specific information to deal with armed crime without relying on the army. The Second Bureau appears to have had an unhealthy though understandable preoccupation with 'Irregular policy', which included purely political developments culminating in the Sinn Féin split and the formation of Fianna Fáil. Furthermore, it was invidious for the national army to engage in large-scale domestic spying through the use of informers and agents, and it was difficult to reconcile such activity with the post-mutiny professionalisation of the army.

Over succeeding years the Gárda Special Branch dealt adequately with the actual menace of militant republicanism and the perceived threat of communism (incidentally spending far less money than had the Second Bureau on informers in the process). They sought and got 'the [Second Bureau] files for "notorieties". De Valéra, etc etc' in the summer of 1926.[32] Presumably such records disappeared, as did many Second Bureau documents concerning informers, civil war courts and the like, just before Fianna Fáil assumed office in 1932.[33] It is not clear whether they continued the surveillance on Fianna Fáil for any length of time, although the volatile Gárda Commissioner Eoin O'Duffy remained unashamedly partisan – he ultimately attempted to interest police and army officers in imposing a military government, naturally with himself as its head, to prevent Fianna Fáil taking office.[34]

After 1926 the army was occasionally called upon to act in support of the Gárdaí, to guard political prisoners and, more disturbingly, to dispense justice through military tribunals for some political crimes. But military intelligence was not allowed to resume its former role, and even its purely military functions were progressively attenuated. For example, it had no attachés abroad, and no means of obtaining relevant information beyond those available to the general public. Its monthly press survey, which at one time had embraced almost seven hundred publications, was run down until it covered only the main Irish and British newspapers, and was stopped altogether in 1930. In 1936 the director of intelligence urged that the Department of External Affairs be told to furnish him with 'an appreciation of the European political situation as it now is ... and ... any information of a military character which it possesses', as 'we receive no information' at all from them.[35] But military intelligence had one considerable asset left when called upon to deal with counter espionage and related security problems in 1938: that was experience. Its director and deputy director, Liam Archer and Dan Bryan, were intelligence veterans of the war of independence and the civil war, and Bryan had been very active against the 'Tobinites' in 1923–4 as well as being the mainspring of the Second Bureau's activities against the republican movement in Dublin up to 1926. These officers were, therefore, exceptionally well qualified for the demanding and sensitive job they were given in 1938–9.

G2's assumption of the primary responsibility for counter-espionage in 1938 seems to have caused little friction. It took some months to sort out 'a variety of difficulties' surrounding the establishment and working of an efficient postal interception service, but once these were dealt with the Department of Justice was generally co-operative in issuing G2 with warrants for postal and later for telephone interception. Since it was agreed at the outset that G2 would work through the Gárdaí rather than establishing its own investigations unit, relations with the police were reasonable. It is, however, unlikely that these arrangements would have been agreed so easily by the Department of Justice had the eventual scope of G2's security work been anticipated.[16] From its modest beginnings as a monitor of limited foreign intelligence activity in Ireland directed against Britain and her friends, once war broke out in Europe G2 quickly became a ubiquitous security organisation. On the declaration of the Emergency the Taoiseach agreed that 'a number of questions affecting the Security of the State, especially in its external aspect', should be the responsibility of G2.[17] This was a very broad charter, and with time it expanded. In furtherance of its original counter-espionage mission, G2 investigated the activities and loyalties of Irish citizens, including some politicians and public servants, as well as of aliens at home and abroad. It advised on the admission of aliens, and on the issue of travel documents for persons wishing to leave Ireland. It spied on Allied and Axis diplomatic missions in Dublin, and wherever possible read or listened to their communications. In addition to postal and telephone interception on a handsome scale, G2 arranged for and controlled the overt postal censorship system. This was designed to prevent the leakage of war information, whether by accident or design, through Ireland. In practice it also yielded a vast amount of miscellaneous information which, while insignificant in security terms, was of considerable background interest and meant that G2 was exceptionally well informed on general conditions and opinion in Ireland and abroad. Although the domestic activities of the republican movement remained primarily the business of the Gárdaí, G2 investigated the IRA's links with Germany and with German agents and Axis sympathisers in Ireland. It also took an interest in communist activity in Ireland, on the grounds that this was heavily influenced by the Soviet Union through the Communist Party of Great Britain. It studied pro-Allied and pro-Axis organisations in Ireland, it watched visiting journalists and, while not formally responsible for press censorship, it had close and amicable contacts with the censorship organisation.

The evidence suggests that throughout the Emergency G2, while it continued to report through army channels on purely military intelligence matters, was virtually an independent agency in the exercise of its counter-espionage and other security functions. This is partly explained by the necessity for extreme secrecy; the reluctance of senior officers to get directly involved in such sensitive questions for fear of political repercussions was probably also a factor.[38] There was a continuous exchange of information with the Gárdaí, whose Special Branch officers did most of the field work against the IRA, aliens, and foreign agents, and the Department of Justice was frequently consulted – the issuing of postal and telephone interception warrants remained in the hands of the Minister for Justice, who usually granted them on request. But G2's relations with the Department of External Affairs were warmer and closer, and the big issues of policy – for example the conduct of security liaison with the British and Americans – were settled in discussion with its senior officials. This benefitted G2 in three important respects: firstly, External Affairs was naturally preoccupied with safeguarding Irish neutrality, analysed every problem in that light, and had no qualms about the means used to maintain it; secondly, the department had never had any such responsibilities before, and unhesitatingly accepted G2's advice on security questions; and thirdly, through the department G2 was effectively working to Mr de Valéra, since he was Minister for External Affairs as well as Taoiseach. This undoubtedly strengthened G2's hand in the delicate matters which it had to handle.

This is not the place for detailed discussion of G2's work during the war, but the range of its activities can be seen from the evidence in the External Affairs records and elsewhere.[39] This includes transcripts of telephone conversations, diplomatic correspondence discreetly opened and copied, partly destroyed documents recovered and restored from fire grates and wastepaper baskets, all manner of information passed to and from MI5, the British security service, and reports on suspect aliens. One covert Axis political organiser, denounced by his own diplomatic mission in 1944, subsequently taught this writer in prep school in the 1960s. Investigations into the doings of a German national revealed that, while entirely innocent of espionage, he was 'carrying on love affairs', 'some of which are conducted by correspondence with ladies he has never met', with a large number of women in Cork, Dublin, Donegal and Germany. 'He appears to have the faculty of playing on women's emotions through correspondence which in due course develops a lascivious tone. He is in fact an expert in sexual psychology', G2 rather grudgingly concluded.[40] External Affairs must have found it extremely useful and occasionally entertaining to know so much about foreign diplomats and journalists, aliens, renegade Irish politicians and officials, and to see the reports of those diplomatic missions too small to protect their mail from prying Irish eyes.

The end of the war in Europe saw a rapid dismantling of the Irish security apparatus built up during the Emergency. G2's director Dan Bryan argued for a substantive peacetime role. On 25 May 1945 he complained to his minister that the system of 'supervision of telephones' operated since 1940 by the Post Office 'solely for this Branch' had been unilaterally discontinued by that department. This worried him because he believed that 'a small [security intelligence] service should be continued and that touch should not again be lost, as happened before the war, with certain internal and external security problems'.[41] When the Department of Justice gave its view that postal warrants would be issued 'solely for the prevention or detection of crime', Bryan sought advice from the Attorney General's office. On that definition, 'several of the more important aspects of a defence security intelligence section could not be covered by postal supervision ... on the grounds that such activities are not criminal in themselves' under Irish law. To make his point he listed various instances of German espionage, subversion and propaganda directed against 'third States such as the United States, France or Great Britain', as well as Soviet use of the IRA against Britain in 1925 and continuing communist subsidy of Irish left-wing organisations, and, perhaps as a balancing item, British recruitment of Irishmen for armed service.[42] As

shown by the experience of this country prior to and during the Emergency one of the problems confronting the state is the necessity of being in a position to prove that the activities in Ireland of certain foreign States, their nationals here, and their Irish friends, are not harmful to other States. The only way in which this can be satisfactorily achieved is for the Defence Security Intelligence Section ... to keep itself fully informed of all such activities.[43]

Bryan's efforts did not achieve a great deal. Telephone and postal 'supervision' were retained, but under the civil authorities. Sporadic interdepartmental dialogue eventually produced agreement in 1954 that the state would always need such services, and four temporary officials so employed since 1939–40 were made permanent civil servants.[44] G2's main reward for a job performed with success and discretion was rapid relegation. The only clear improvement, in comparison with the pre-Emergency doldrums, was that it retained some responsibility for counter-espionage. In so far as External Affairs records up to 1961 are any guide, this mainly meant the monitoring of third country use of Ireland as a base for espionage and subversion against Britain and her allies. G2 also continued to liaise with foreign intelligence and security services, principally those of Britain and America.[45]

Comparison of G2's Emergency activities with those of the Second Bureau in the mid-1920s brings out a number of differences. Firstly, during the Emergency G2 operated by and large in tandem with the police, not in competition with them, and there was an equilibrium of sorts between them. There was a reasonably clear division of responsibilities, and there was plenty to occupy both organisations. Thus G2 did not attempt to penetrate the IRA on a significant scale, and did not create its own networks of informers on domestic political affairs, mainly relying on the Gárdaí. On the other hand, through postal censorship, through telephone tapping, through liaison with foreign security agencies and through its own officers serving with units throughout Ireland it had considerable sources of information of its own. Secondly, because of the external military threat to Irish neutrality G2 had onerous military tasks in addition to its security and counter-espionage functions, and this probably absorbed energies that might otherwise have gone into building an even larger security and intelligence empire. In the 1920s there had been no realistic external military threat at all to dilute Second Bureau absorption in domestic intelligence work. Thirdly, while generally left to get on with its work G2 operated along lines laid down by ministers, in defence of the policy of neutrality on which there was an overwhelming political consensus. Fourthly, between 1938 and 1945 it was impossible in practice to separate questions of internal security and domestic politics from external defence. In contrast to the mid-1920s, G2's extensive intelligence and internal security activities during the Emergency derived from the army's formal professional role as defender of the state from external threat; paradoxically, however, its great success stemmed largely from the experience and judgement of officers who had learned their trade in the internecine violence and political intrigue of the state's first years.

NOTES

1   See the discussion of the principle of civil supremacy in S.E. Finer, *The Man on Horseback: The Role of the Military in Politics* (2nd ed., London 1988), pp. 22–6.

2   Basil Chubb, *The Constitution and Constitutional Change in Ireland* (Dublin 1978), p. 30; Michael Gallagher, 'The Presidency of the Republic of Ireland: Implications of the "Donegan affair"', *Parliamentary Affairs* 30 (1977), p. 378; *The Irish Times*, 1 Nov. 1990.

3   F.H. Hinsley and C.A.G. Simpkins, *Security and Counter-Intelligence* (London 1990), pp.

1–25, vol. 4 of *British Intelligence in the Second World War* (5 vols, London 1979–90).

4   Bryan tapes, p. 2 (in the possession of this writer). This writer interviewed Colonel Bryan on a number of occasions between 1983 and 1985. In addition, in 1984 Colonel Bryan began tape-recording his memoirs, and later gave this writer fifty three pages of transcripts of these recordings.

5   'Fundamental Factors Affecting Irish Defence Policy', a document marked 'G2/0057, dated May, 1936', with private secretary, Minister for Defence, to private secretary, Minister for Finance, 23 May 1936 (U[niversity] C[ollege] D[ublin Archives], Sean MacEntee papers, P67/191, pp. 58–9, 68, 79–80); Bryan tapes, p. 2.

6   Chief of staff to Minister for Defence, 22 Sept. 1936 (MacEntee papers, P67/191).

7   Copy of memorandum on 'Defence Security Intelligence', with copy of Bryan (as director of intelligence) to Minister for Defence, 21 June 1945 (N[ational] A[rchives, Dublin], D[epartment of] E[xternal] A[ffairs], A8/1).

8   Conor Brady, *Guardians of the Peace* (Dublin 1974), pp. 202–5.

9   Extract from Civic Guard report, n.d., with Director of Intelligence to Chief of Staff, 30 Nov. 1923 (M[ilitary] A[rchives], 'Intelligence Branch' file on 'Co-operation with Civic Guards').

10   Eunan O'Halpin, 'Intelligence and Security in Ireland, 1922–45', *Intelligence and National Security*, v, No. 1 (Jan. 1990), pp. 52–3; Army Finance Office to Department of Finance, 25 Sept. 1923 (NA, FIN1 747/104).

11   Unsigned report, 3 July; report by intelligence officer, Dublin District, 27 July 1923; ibid, 14 Jan. 1924 (MA, 'Intelligence' file on 'Owens, Joe, York Road, Dun Laoghaire. Activities of').

12   'Report on Working of Intelligence Department for Period: 1st October 1924/31st December 1925', undated, in file on 'Annual Report 21/1/26–27' (MA, cited hereafter as '1925 Report').

13   Ibid.

14   J.J. Lee, *Ireland 1912–1985: Politics and Society* (Cambridge 1989), pp. 101–3; quoted in O'Halpin, 'Intelligence and Security', p. 54.

15   '1925 Report'; O'Halpin, Ibid., pp. 54–5.

16   '1925 Report'.

17   Ibid.

18   Ibid.; O'Halpin, 'Intelligence and Security', pp. 61–2.

19   The deciphered text of IRA documents on this are in UCD, Desmond Fitzgerald papers, P80/869; Dan Bryan (Director of Intelligence) to Philip O'Donoughue SC, Office of the Attorney General, 21 June 1945, with Bryan to Joseph Walshe, Secretary, Department of External Affairs (NA, DEA A8/1).

20   '1925 Report'.

21   Report dated 21 Nov. 1924 (UCD Fitzgerald papers, with many others in P80/847).

22   Director of intelligence to chief of staff, 19 Nov. 1925 (ibid, P80/849).

23   Bryan tapes, p. 44; note initialled by Comdt Brennan Whitmore, 11 Oct. 1924 (as in note 9 above).

24   Memorandum by Captain J.J. Feeney to director of intelligence, 13 Aug. 1925 (ibid.).

25   Memorandum by director of intelligence on 'Co-operation of Military and Civil Forces', 14 Sept. 1925 (ibid.).

26   Chief of staff to Minister for Defence, 8 Dec. 1925 (ibid.).

27   Director of intelligence to assistant chief of staff, 3 Feb. 1926 (ibid.).

28 Chief of staff to commissioner, Gárda Siochána, 6 Aug.; director of intelligence to chief of staff, 18 Oct. 1927 (ibid.).

29 Extract from Eastern Command intelligence report for July 1926 (UCD Desmond Fitzgerald papers, P80/847); 'Annual Report on Working of Second Bureau, General Staff Period 1st January, '26, to 31st December, '26' (MA, cited hereafter as '1926 Report').

30 O'Halpin, 'Intelligence and Security', p. 55.

31 Annual report by Commandant Dan Bryan, Eastern Command intelligence officer, 31 Jan. 1927, in file on 'Annual Report 21/1/26–27' (as in note 12 above).

32 Chief superintendent Neligan, Gárda Síochána, to director of intelligence, 19 June 1926 (as in note 9 above). The overall amount spent each year from the secret service vote can be found in the annual *Appropriation Accounts*. This writer also has a copy of a Department of Finance document giving a breakdown of spending by each department from 1922 to 1936. In 1928/9 the army's share was just £12. For the history of the secret service vote, see Eunan O'Halpin, 'The Secret Service Vote and Ireland, 1868–1922', *Irish Historical Studies*, xxiii, no. 92 (Nov. 1983), pp. 348–52.

33 O'Halpin, 'Intelligence and Security', p. 55.

34 Lee, *Ireland 1912–1985*, pp. 175–6; Brady, *Guardians of the Peace*, pp. 167–9.

35 Report by Captain J.A. Power, 4 Jan. 1927 (as in note 29 above); unsigned MS note, 15 Apr. 1930, Desmond Fitzgerald papers, P80/1043.

36 Memorandum on 'Defence Security Intelligence', with director of intelligence to Minister for Defence, 21 June 1945, as in note 7 above; O'Halpin, 'Intelligence and Security', p. 66; Bryan tapes pp. 47–9; interview with Colonel Bryan, Jan. 1983. Colonel Bryan recalled encountering severe difficulties with the Department of Justice once Peter Berry took charge of security matters in 1941.

37 Memorandum on 'Defence Security Intelligence' (as in note 36 above).

38 Interview with Colonel Bryan, July 1983.

39 In addition to External Affairs files in the National Archives, and G2 files in the Military Archives (not all of which are open to research), there is a good deal of material in the papers of Dr Richard Hayes in the National Library of Ireland (Ms.22983 and 22984).

40 Minute by F.H. Boland, Department of External Affairs, 3 Jan. 1944, and attached documents (NA, DEA A63); MS addition by Colonel Bryan to TS 'notes on activities' of General Eoin O'Duffy, 19 Apr. 1943 (DEA A62); Dan Bryan to F.H. Boland, 12 Jan. 1944 (ibid.); TS report, undated, July 1944 (MA, G2/2457).

41 Copy of Bryan to Minister for Defence, with Bryan to Walshe, both 25 May 1945 (NA, DEA A8/1).

42 Bryan to Minister for Defence, and Bryan to Philip O'Donoughue, both 21 June 1945, with Bryan to Walshe, 25 June 1945 (ibid.).

43 Memorandum on 'Defence Security Organisation' (as in note 7 above).

44 Documents on this area are in 'Assistant Inspectors Post Special Duties, 1939–54' (NA, Department of Communications, H3013/35).

45 Interview with Colonel Bryan, July 1983; interview with a former CIA officer who worked on liaison with G2 while attached to the American embassy in London during the 1950s and 1960s, Cambridge, July 1988. See also Allen Dulles, Director, Central Intelligence Agency, to the Minister for External Affairs, 26 Feb. 1955, seeking a 're-establishment of liaison between our respective intelligence services' (NA, DEA A60/1).

HITLER'S HALTBEFEHL BEFORE DUNKIRK
A MILITARY OR POLITICAL DECISION?

Jean Stengers

Hitler's order of 24 May 1940 to halt the German Panzers as they were approaching Dunkirk – the celebrated *Haltbefehl* – remains one of the riddles of the Second World War. It is a riddle of importance, for without that order the course of the war could have been changed.

The facts are not disputed.

On the eve of 24 May 1940, two German Army groups were gaining ground against the British, French and Belgian forces encircled in Flanders: Army Group B, under General von Bock, which had crossed Belgium, and Army Group A, under General von Rundstedt, which had swept through Northern France from Sedan to the Channel. The main German attack scheduled for the next day was not to be a frontal attack against Dunkirk (until the re-embarkment began there, as the Germans had not envisaged the possibility of such a large-scale operation, Dunkirk was not a high priority). The plan was for a second Sedan manoeuvre: a breakthrough in the very centre of the Allied forces so as to divide them into two powerless halves. General Halder, the Chief of the Army General Staff, had planned that attack along the line Armentières–Ypres–Dixmude–Ostend (see map overleaf).[1] That decisive blow was to lead to the total destruction of the enemy.

On the morning of 24 May, Hitler arrived at Charleville to confer with General von Brauchitsch, the Commander in Chief of the Army, at his headquarters. Just after 12.30, the order was given to Army Group A: all armoured forces must stop on the line Lens–Béthune–Saint-Omer–Gravelines (see map). It was an order given first by telephone – which meant of immediate execution – and afterwards confirmed in writing.[2] Accompanying that *Haltbefehl* was a directive for the German Air Force: it was the Luftwaffe which had henceforth the main responsibility for destroying the encircled enemy.[3]

The *Haltbefehl* was cancelled only on 26 May. After the offensive was again allowed, it still took some time for the armoured divisions to prepare for it.

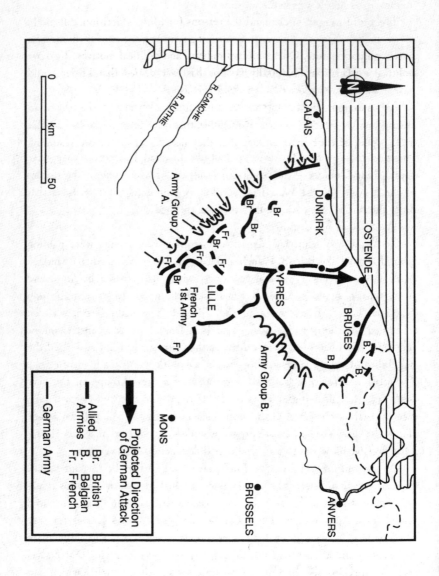

The situation on 24 May 1940

In the meantime, the British and French forces had been permitted to reorganise their own defence. They could vigorously resist the new attack. So the 'miracle of Dunkirk' was made possible.

The problem is to understand the reasons for Hitler's decision. All in all, no fewer than seven explanations have been given for it. Four of them are of a purely military nature. One combines military and political motives. Two are political explanations. Naturally no one has ever denied that Hitler could have obeyed two or more motives simultaneously.

Let us examine the military explanations first . They are: (1) the fear of an enemy counter-attack; (2) the consideration that the Panzer divisions had lost a large part of their material and that they needed repairs before resuming their operations; (3) the fear of the Flanders marshy grounds, where Panzers would have a particularly difficult task; (4) the desire to spare the Panzers for the great final offensive against France.

Motives (1) and (2) are hypotheses which find no foundation in contemporary sources or in serious testimonies.

As regards (1), there is no trace whatever of such a fear, which could have led to prudence – at least at the time when the decision was taken on 24 May. The general climate at that time in the German high command was one of robust optimism: the enemy seemed to be no longer capable of any dangerous reaction.[4]

As for (2), it is clear that the losses of material in the German armoured divisions had been high, but it is also clear that the generals commanding these units were generally eager to go on with the attack and were profoundly disappointed when they were stopped.[5] There is no indication of a pessimistic analysis in the case of Hitler himself.

Motive (3) is alluded to in a 'three-star' source: the diary of Halder. The watery Flemish grounds, he writes on 25 May, are described as unsuitable for tanks.[6] Halder expanded on the subject in a letter he sent years later, in 1957, to the American journalist and historian William Shirer. He wrote then: 'According to my still quite lively memory, Hitler, in our talks at the time, supported his reasons for the stop order with two main lines of thought. The first were military reasons: the unsuitable nature of the terrain for tanks, the resulting high losses which would weaken the impending attack on the rest of France, and so on'.[7] In May 1940, according to his diary, Halder seems to have considered that reason mainly as a pretext. But, pretext or not, the reason was only valid for an attack in the direction of Dunkirk, and in the coastal region. It had no relevance for the great Halder manoeuvre, for the thrust towards Ypres and Ostend: marshy grounds would not have embarrassed the tanks there. And yet that was the main attack which was stopped. Motive (3) was thus at best a very partial explanation.

Motive (4), on the contrary, was certainly at the time a good one for the *Haltbefehl* as a whole. It is underlined in contemporary documents.[8] As far as we understand Hitler, it must have been a part of his reasoning on 24 May.

A fifth motive has been alleged, which combines military and political reasons. By leaving the final work of destruction mainly to the Luftwaffe, Hitler was sparing his Panzers, he was sparing his field forces, but he was also doing a favour to Goering and to a branch of the Wehrmacht which was politically more devoted to him than the army. Goering had strongly and repeatedly insisted on that point:[9] if the glory of victory 'could be claimed exclusively by the army generals, the prestige of the Führer in the German homeland would be damaged'.[10] The Luftwaffe must be allotted a large part of the victory, by striking the final blow. This fifth motive, too, is well documented, and must be taken into account.

Motives (6) and (7) are the purely political motives which have been put forward.

Motive (6) has enjoyed, and still enjoys, a degree of popularity, but it is preposterous: it is the idea that Hitler wanted to spare Britain so as to facilitate a peace settlement. That explanation made its first appearance only after the war,[11] and it was advanced by some German generals. General Blumentritt was apparently the first to speak in that way in 1945, in his conversations with Liddell Hart. 'He felt', wrote Liddell Hart (and the verb 'feel' is a significant one for the nature of his testimony), 'that the 'halt' ... was a part of a political scheme to make peace easier to reach. If the British Army had been captured at Dunkirk, the British people might have felt that their honour had suffered a stain which they must wipe out. By letting it escape Hitler hoped to conciliate them'.[12] Blumentritt's 'feeling' has been echoed in many parts. It is all pure nonsense. We very clearly perceive Hitler's intentions on 24 May and in the following days, and as regards the British Army, they may be summed up in one word *Vernichtung*, annihilation.[13] In his general directive of 24 May, issued at the same time as the *Haltbefehl*, Hitler felt so benign towards England that, apart from the *Vernichtung* of the encircled British forces, he authorised the Luftwaffe 'to attack the English homeland in the fullest manner'.[14] *Exeunt* Blumentritt and the like.

But Halder pointed to another political motive, motive (7), which has been generally overlooked, and which merits every consideration. Just one day after the *Haltbefehl*, on 25 May, Halder noted in his diary that the order has had a 'political objective': 'The political command has come to the idea that the final decisive battle must not be fought on the soil of the Flemish people, but rather in Northern France'.[15]

The main reason why that reference to Hitler personally – the 'political command' – has in most cases not been taken seriously, is that historians have been obsessed with Dunkirk, and have looked at Dunkirk as the central issue of the whole battle. Refraining from an attack on Dunkirk because it would have taken place on 'the soil of the Flemish people' makes little sense. But this is to forget where the main final offensive was to be launched, i.e. in the direction of Ypres and Ostend: there, in the most proper sense of the word, was the Flemish soil.

William Shirer was a little puzzled by that entry in Halder's diary, and he asked Halder himself for some explanation. Halder, in a letter of 19 July 1957, gave ample confirmation of what he had noted in his diary. After a reference to Hitler's military motives (those we have quoted before), he recalled that the Führer

cited a second reason which he knew that we, as soldiers, could not argue against since it was political and not military. This second reason was that for political reasons he did not want the decisive final battle, which inevitably would cause great damage to the population, to take place in territory inhabited by the Flemish people. He had the intention, he said, of making an independent National Socialist region out of the territory inhabited by the German-descended Flemish, thereby binding them close to Germany. His supporters on Flemish soil had been active in this direction for a long time; he had promised them to keep their land free from the damage of war. If he did not keep this promise now, their confidence in him would be severely damaged. That would be a political disadvantage for Germany which he, as the politically responsible leader, must avoid.[16]

Are these recollections of Halder plausible? Undoubtedly so, for such a political interpretation of the *Haltbefehl* is quite compatible with a series of decisions taken by Hitler, or around Hitler, between April and July 1940. Compatibility is the word that best describes these various interventions.

To begin with: on 10 April 1940, directives were issued concerning the future Belgian prisoners of war. Flemish and Walloon prisoners should be segregated. This evidently implied better treatment for the Flemings.[17]

On 17 May, one week after the beginning of the campaign in Belgium, the German press was officially asked (and Goebbels naturally knew what Hitler thought) to adopt a specific language as regards the Flemish population and to abstain from any offence against them.[18] At the end of the month, German newspapers were reported as publishing maps underlining the linguistic division of Belgium, as well as insisting upon the unnatural character of the combination of the two elements of the Belgian population.[19]

In the military field, on 23 May, orders were given by the High Command

of the Army 'to spare as far as possible the great cities of the Flemish country'. The 'preservation' of these cities, it is said, is 'of value for the *political command*' (my emphasis).²⁰ Among all the others, this is really the key-text; it
dates from one day before the *Haltbefehl.*

The capitulation of the Belgian army, on 28 May, made it necessary to
settle definitely the question of the prisoners of war. The personal orders of
the Führer were: to separate the Flemings and the Walloons, to liberate the
Flemings, while the Walloons would probably be taken away to Germany.²¹
These orders were reiterated in a more precise form – once again by Hitler's
personal decision – on 5 June.²² Nevertheless it would take some weeks and
even months before they could be fully implemented:²³ the word of the
Führer was one thing, the administrative machine was another.

At the beginning of June, a high officer of the OKW revealed what Hitler
envisaged for the political future of Belgium: the Flemish territory to be
placed under a *Reichskommissar*, while the Walloon part of the country would
retain a military government.²⁴ This solution would not actually be adopted,
but the mood of the Führer is in itself very significant: to give the Flemish a
Reichskommissar was, in his view of things, to give them a mark of favour.

Favour was also reflected on 19 June at Goering's headquarters: cattle
should be requisitioned in Belgium and northern France and sent to Germany, but this was not to apply to the cattle of the Flemings.²⁵

Finally, we come to the general directive given by Hitler on 14 July: pending a final decision about the political future of Belgium, the Führer wanted
'every possible favour for the Flemings, including the repatriation of the
Flemish prisoners of war ... For the Walloons, no advantage of any kind'.²⁶
This, once again, is a real key.

How can we explain this special sympathy for the Flemings? Three main
reasons can be put forward.

(1) Hitler had allies among the Flemish extremists, and he had made them
promises which he intended to keep. In his letter of 1957 to William Shirer,
Halder was very specific about that. Too specific, maybe. After seventeen
years, his memory might easily have lapsed.

What we know for certain is that even before May 1940 a tiny number of
Flemings had engaged in work for Nazi Germany. They were the members of
a clandestine military organisation, the *Militaire Organisatie*, whose activities
were supervised by the leader of the Flemish Nationalist Party (*VNV, Vlaamsch
Nationaal Verbond*), Staf de Clercq, and aimed at demoralising Flemish soldiers
in the army, and inciting them to offer no resistance to the Germans.²⁷ The
Militaire Organisatie had close contacts with the *Abwehr*, and an *Abwehr* envoy in

Belgium seems to have been very happy about the development of its work.[28] Pamphlets to be distributed to Flemish soldiers were printed in Germany and brought to Belgium so as to be of immediate use once the hostilities began.[29] Two problems, however, are not completely solved:

(a) Did the *Militaire Organisatie* play a really effective part during the May 1940 campaign? Staf de Clercq publicly boasted later that it had saved the lives of thousands of German soldiers.[30] This may have been just a boast, to show to the Germans how helpful he had been. A close scrutiny of the events, and of the activities of members of the *Militaire Organisatie*, rather tends to reduce to very little their impact on the May 1940 campaign. It is true that, especially at the end of the campaign, some Flemish units offered little, or no resistance, but they probably did not need any propaganda to be demoralised.

(b) More important: did Hitler know about the *Militaire Organisatie*? There is no proof that he did, but one is inclined to think that this was the kind of thing about which he was informed. The preparation of the offensive in the West was of such vital importance that the problem of the morale of the Belgian army must have been one of his preoccupations. This could explain Halder's allusion in his 1957 letter.

(2) Hitler thought that the general attitude of the Flemings towards Germany was different from that of the Walloons. That was his emphasis when he spoke in October 1940 with Princess Marie-José of Italy, the sister of King Leopold III, and in November 1940 to King Leopold himself.[31] Among Flemings, he told the King, there had been a sympathetic attitude. Even if this analysis was a wrong one – and it largely was – it certainly formed part of Hitler's reasoning, or rather emotional reaction.

The future, in that respect, would sadly disillusion Hitler. After a few months or a year of German occupation in Belgium, it must have been obvious even to those who had had preconceived ideas that, apart from some collaborators in the south as well as in the north of the country, Germany had to cope with the same hostility of the mass of the population in both areas. This may explain why Hitler did not reiterate his July 1940 directive: all favours to the Flemings, none to the Walloons.

(3) Preconceived ideas, in the case of Hitler, were certainly linked with his conceptions of race. The Germanic race, for him, was the paramount one. His dream was that of a great Germanic empire dominated by the Germans, but to which other parts of the Germanic race – the Flemings among others – would belong. The best interpreter of Hitler's views is probably Gebhardt, a close collaborator of Himmler, when he explained in Brussels that the prob-

lem of the prisoners of war must be solved according to 'racial' principles.[32] Once again, there is a perfect coherence in this attitude. The Führer had already ordered on 9 May the liberation of the Norwegian prisoners of war, and the same measure was taken on 1 June for the Dutch, members of that Dutch people which was, the Führer's proclamation said, 'a parent of the Germans by race'.[33] Such were the first delineaments of the construction of the great Germanic Reich.

One may wonder whether yet another motive was not at work in Hitler's mind: during the First World War, he had spent long months in Flanders. It does not seem to have been an unhappy period of his life.[34] Did he have a kind of nostalgia, a degree of attachment to the country? This, one must confess, is pure psycho-history, which actually leads to no more than an arbitrary hypothesis.

At the end of this analysis, we are left with three apparently valid motives for the *Haltbefehl*: number 4 (to spare the armour for the offensive against France); number 5 (to leave the final victory to the Luftwaffe, and to spare at the same time the ground forces) and number 7 (to spare the Flemings and the Flemish country). How did these motives combine in Hitler's mind? There is no way to solve the riddle of a man's mind. But whatever the importance of the different motives, there are good reasons to believe that motive number 7 had real weight.

Of one thing we may be practically certain: without the *Halfbefehl*, Halder's big thrust towards Ypres and Ostend would have had every chance of success. Halder himself was quite confident: we have no enemy in front of us, he noted in his diary.[35] This is confirmed by the British themselves: in a report about the campaign, General Alan Brooke mentioned the luck they had of seizing on 25 May in the documents of a German staff officer the plans for the German attack towards Ypres, and he adds: 'We had at that time nothing on the Ypres–Comines Canal except one brigade'.[36]

With the success of that German offensive, there would have been no Dunkirk. It would have meant the loss of the greatest part of Britain's field army, of 'practically all our trained soldiers', General Ironside wrote with some exaggeration,[37] the loss at any rate of the best trained officers of the army. The consequences could have been immense.

If the *Haltbefehl* deprived Hitler of a decisive victory in the West, and if our explanation for the *Haltbefehl* is valid, we may see in Hitler's destiny a kind of poetic justice. Race was a driving idea in Hitler's mind. The idea was at the heart of some of his greatest impulses. But race at the same time contributed to his defeat both on the Western and the Eastern front. In the East, such was his hatred and contempt for the Slav *Untermenschen* that he rejected with scorn in 1941 the thousands and even hundreds thousands of them who – like so many Ukrainians – would have been ready to collaborate with Germany, because they hated the Soviet regime. With their help, the best experts now agree, Hitler could have been victorious. But race was his doom. Because he wanted to spare his Flemish racial brethren, he lost in the West. Because he utterly despised the Slavs, he lost in the East.

To this note of ironic justice, we may add, as a final touch, one of high comedy. General Van Overstraeten, the military adviser of Leopold III, could take decisions in the name of the King. On 22 May 1940, as he observed the movements of the German army, he came to the conclusion that perhaps the objective of the Germans was to encircle the Belgian forces, but not to attack and destroy them. If this was the case, the Germans had to be warned of the point where to stop. So Van Overstraeten sent a staff officer to the Franco-Belgian frontier ordering him to erect barricades with posters bearing the inscription: 'Ici Belgique'.[8] This may sound hilarious. But with what we now know of Hitler, it would not have been so preposterous after all to have put on the barricades the inscription: 'Hier Flandern – Here Flanders'.

NOTES

1   Documents 7, 8 and 15.

2   Documents 9, 10 and 12. The testimonies of German generals, in their conversations in 1945 with Liddell Hart, are also worth noting: 'Field-Marshal von Runstedt and General Blumentritt (in May 1940 a member of Rundstedt's General Staff) gave me their respective accounts of how the order had come to Army Group A – transmitted over the telephone by Colonel von Greiffenberg at OKH. Blumentritt said that he had himself taken the telephone call.' See B.H. Liddell Hart, *The Other Side of the Hill* (revised edn, London 1951), p. 188. The order was immediately repeated to the chief units of Army Group A. A message sent *en clair* at 12.42 p.m. ordering to discontinue the attack on the line Dunkirk–Hazebrouck–Merville (the sector nearest to the coast) was intercepted by the British (see Winston Churchill, *The Second World War*, vol. ii, *Their finest hour* (London 1971), p. 68, who gives the British time of 11.42 a.m.; also in Brian Bond (ed.), *Chief of Staff. The diaries of Lieutenant-General Sir Henry Pownall*, vol. i (London 1972), pp. 337–8.

3   Documents 6, 11 and 13.

4   Documents 12 and 13.

5   'Die meisten drängten zum weiteren Angriff', Hans-Adolf Jacobsen points out in his *Dünkirchen. Ein Beitrag zur Geschichte des Westfeldzuges 1940* (Neckargemünd 1958), p. 83. Particularly revealing in this respect are the entries for 23 May of the *Kriegstagebücher* of XXI Corps (General Reinhardt) and of the 6th Armoured Division (see H. Meier-Welcker, 'Der Entschluss zum Anhalten der deutschen Panzertruppen in Flandern 1940', *Vierteljahrshefte für Zeitgeschichte* (July 1954), pp. 277, 277–8, and H.-A. Jacobsen, *Dünkirchen*, pp. 83–4 and 84).

6 Document 15.

7 Letter of 19 July 1957 quoted in William L. Shirer, *The Rise and Fall of the Third Reich* (Crest Printing, New York 1962), p. 966. General Kleist told Liddell Hart in 1945 that when meeting Hitler a few days later, he 'ventured to remark that a great opportunity had been lost of reaching Dunkirk before the British escaped. Hitler replied: "That may be so. But I did not want to send the tanks into the Flanders marshes."' See B.H. Liddell Hart, *The Other Side of the Hill* (London 1948), p. 140.

8 Document 9. 'Man will sie (die Panzerverbände) wohl für spätere Aufgaben schonen' (Diary of General von Bock, Commander of Army Group B, entry for 24 May 1940; H.-A. Jacobsen (ed.), *Dokumente zum Westfeldzug 1940* (Göttingen 1960), p. 77. 'Die Panzer zu schonen' (*Kriegstagebuch* of Army Group A, entry for 25 May 1940; *Dokumente zum Westfeldzug*, p. 80). Rundstedt testified before a Commission of the International Military Tribunal at Nuremberg in 1946: When 'Hitler stopped the armored component of my army group ... he stated that the reason for this was that in the second phase of the French campaign ... that he did not want to wreck those armored divisions needlessly' (quoted by Telford Taylor, *The March of Conquest. The German Victories in Western Europe, 1940* [London 1959], p. 385 n. 88).

9 Document 6.

10 Letter of Halder of 19 July 1957 quoted by Shirer, *Rise and Fall*, p. 965. Halder does not give here a direct testimony but repeats the accounts given by the Luftwaffe Generals Milch and Kesselring. See also Franz Halder, *Hitler seigneur de la guerre*, French trans., R. Jouan and L. Rougier (eds) (Paris 1950), p. 60.

11 One quotes sometimes, however, a passage of Hitler's so-called 'political testament', which is the name given to the notes of his conversations in February and April 1945 allegedly taken by Martin Bormann. The comment he is said to have made on 26 February is: 'Churchill hat die Grosszügigkeit und Ritterlichkeit nicht zu schätzen gewußt, die ich durch Vermeidung des Äussersten England gegenüber hundertfach bewiesen habe. *Absichtlich habe ich die flüchtenden Briten bei Dünkirchen geschont*' [our emphasis] (*Hitlers Politisches Testament. Die Bormann Diktate von Februar und April 1945* [Hamburg 1981], p. 113). Unfortunately there are very good reasons to believe that the 'political testament' is a complete fabrication.

12 Liddell Hart, *The Other Side of the Hill* (revised edn), p. 201.

13 Documents 6, 11 and 13.

14 Document 11.

15 Document 15.

16 Shirer, *Rise and Fall*, p. 966.

17 Document 3.

18 Document 5.

19 Document 18.

20 Document 7.

21 Document 17.

22 Document 21.

23 Strangely, in the second half of September 1940, among the prisoners of war detained in Germany, there was still a majority of Flemings (see Kiewitz to Lammers, 21 September 1940, in *Akten zur Deutschen Auswärtigen Politik, 1918–1945*, series D, 1937–45, vol. xi/1 [Bonn 1964], p. 125). Tens of thousands of them were repatriated and liberated only after that date. See on the whole question, G. Hautecler, 'L'Origine et le nombre des prisonniers

de guerre belges, 1940–1945', *Revue Internationale d'Histoire Militaire* (1970), and L. De Vos and J.P. Descan, 'De Taalverhoudingen bij de Belgische krijgsgevangenen tijdens de Tweede Wereldoorlog', *Wetenschappelijke Tijdingen*, vol. xlviii (1989).

24   Document 19.

25   Document 22.

26   Document 23.

27   On the general problem of the *Militaire Organisatie*, see Maurice De Wilde, *De Kollaboratie*, vol. i (Antwerp 1985), pp. 63–73, and Bruno De Wever, *Staf De Clercq* (Brussels 1989), pp. 137–55.

28   Documents 2 and 4. On the reliability of these entries of the Kriegstagebuch of Abwehr II, see Etienne Verhoeyen in *Revue Belge de Philologie et d'Histoire*, vol. lxviii (1990), pp. 1049–50. Also L. Papeleux, 'L'Activité de l'Abwehr II en Belgique', *La Vie Wallonne*, vol. xliii (1969), pp. 99–107.

29   Document 1. See Document 16 for an example of these pamphlets.

30   Document 25.

31   Documents 24 and 26.

32   Document 20.

33   Max Domarus, *Hitler. Reden und Proklamationen, 1932–1945*, vol. ii, 1st part (Munich 1965), pp. 1501, 1519.

34   This tentative remark naturally excludes what he suffered on the battlefield, and especially during the gas attack of October 1918. In the large literature on the subject, see for instance Werner Maser, *Hitler, Adolf*, French transl. (Paris 1973), chapt. 4, or Henri Bourgeois, 'Hitler et la région de Comines–Warneton en 1914–1918 et en 1940', *Mémoires de la Société d'Histoire de Comines–Warneton et de la région*, vol. xvii (1987). The artist Hitler's works during his stays in Flanders – which could easily be an inspiration for psycho-historians – are in *Adolf Hitler als Maler und Zeichner. Ein Werkkatalog*, Billy F. Price (ed.) (Zug 1983)

35   Document 13.

36   Document 14. On this seizure of German staff documents, see L.F. Ellis, *The War in France and Flanders, 1939–1940* (London 1953), pp. 148–9; J.R. Colville, *Man of Valour. The Life of Field-Marshal the Viscount Gort* (London 1972), p. 217; F.H. Hinsley, *British Intelligence in the Second World War*, vol. i (London 1979), pp. 143, 161–2. Brooke himself has been more explicit in an autobiographical note published by Arthur Bryant, *The Turn of the Tide, 1939–1943* (London 1957), pp. 122–5. There is some discussion – but this does not bear upon our subject – about the relationship between the seizure of the documents and Gort's later decisions: see David Irving, *Churchill's War*, vol. i (London 1989), pp. 292–3.

37   *Time Unguarded: the Ironside Diaries, 1937–1940*, R. MacLeod and D. Kelly (eds) (London 1962), p. 333.

38   Diary of Commandant Jean Verweyen, quoted in Henri Bernard, *Panorama d'une défaite. Bataille de Belgique–Dunkerque, 10 mai–4 Juin 1940* (Paris 1984), pp. 122–3, and General Van Overstraeten, *Albert I–Léopold III. Vingt ans de politique militaire belge, 1920–1940* (Brussels 1949) p. 663.

DOCUMENTS

1.   *Kriegstagebuch* of *Abwehr II,* January 1940, (Institut für Zeitgeschichte, München)

   10.1.40: Der Druck des für Belgien vorbereiteten Zersetzungsmaterials ist in Auftrag gegeben.

13.1.40: Beginn der Drucklegung des für Belgien und Holland vorgesehenen Flugblatt-materials (Reichsdruckerei).

16.1.40: Dr Heller, Mitarbeiter Referat 1 West, ist nach der Überbringung von Propaganda – und S-Material nach Brüssel in der Nacht vom 15. zum 16. Januar wieder auf deutschem Gebiet eingetroffen.

2. *Kriegstagebuch* of *Abwehr II*, 14 March 1940, (Institut für Zeitgeschichte, München)

14.3.40: Dr Scheuermann (Beauftragter des Referats I) von seiner Reise nach Belgien zurück. Er hatte Besprechungen mit dem Vlamenführer Staf de Clercq und andere V-Leute. Ergebnis: ... Im Verlauf der letzten drei Monate ist nach Angaben eines als zuverläßig beurteilten und speziell in dieser Hinsicht angesetzten V-Mannes ein fühlbarer Aufschwung der vlamischen Militärorganisation (der sogenannten Regimentsvereine) eingetroffen.

3. Instructions of the *Oberkommando des Heeres* (*OKH*), 10 April 1940, (see W. Wagner, *Belgien in der deutschen Politik während des Zweiten Weltkrieges* [Boppard 1974], p. 160)

Belgische Kriegsgefangene sind möglichst bereits während ihrer Rückführung nach Flamen und Wallonen zu trennen, und zwar in der Regel danach, ob sie zu einem flämischen oder wallonischen Regiment gehört haben. Eine Ausnahme hiervon ist für die in Brüssel beheimateten Gefangenen zu machen. Soweit diese von Hause aus flämisch sprechen, sind sie ohne Rücksicht auf ihre bisherige Regimentszugehörigkeit ebenfalls unter die flämischen Kriegsgefangenen einzureihen.

4. *Kriegstagebuch* of *Abwehr II*, May 1940, (Institut für Zeitgeschichte, München)

3.5.40: Verschiedene Meldungen der letzten Wochen lassen erkennen, daß bei den flämischen Truppenteilen der belgischen Armee ernsthafte Schwierigkeiten und Gehorsamsverweigerungen wiederholt vorgekommen sind. Diese Tatsache dürfte auf die verstärkte Arbeit der sogenannten Regimentsvereine, d.h. der Militärorganisation der VNV zurückzuführen sein.

9.5.40: Bericht eines aus Belgien zurückgekehrten Beauftragten des Referates 1 West, Dr Scheuermann, über seine Reise nach Belgien und Holland in der Zeit vom 22.4 bis 1.5.40. Dr Scheuermann führte Besprechungen mit verschiedenen V-Leuten. Besonders erwähnenswert:

1) Staf de Clercq, Führer des VNV. Staf de Clercq bemüht sich, die Ausbreitung der Militärorganisationen und Insurgierung des belgischen Heeres zu steigern ...

14.5.40: Ich teile fernmündlich mit, daß der Chef OKW im Führerhauptquartier seine Anerkennung für die Tätigkeit der Abwehr II im Zusammenhang mit dem Beginn der Operationen am 10. Mai ausgeschprochen hat.

31.5.40: Aus den Feststellungen, die Major Marwede in Brüssel gemeinsam mit flämischen V-Leuten getroffen hat, geht schon jetzt hervor, daß die militärischen Organisationen (die sogenannten Regiments-Vereinigungen) des VNV erheblich dazu beigetragen haben, durch ihre Beeinflußung der flämischen Truppen die Stimmung im belgischen Heer weitesgehend herabzumindern und in den Flamen den Wunsch zu erwecken, sich Kampfhandlungen nach Möglichkeit zu entziehen.

5. Instructions to the German press: Press conference of 17 May 1940, (Koblenz, Bundesarchiv, ZSg. 101, Sammlung Branner)

Anweisung Nr 85.

Die Volksdeutsche Mittelstelle bittet im Einverständnis mit dem Prop. Min., die Volkstumsfragen in Holland und Belgien im deutschen Sinne langsam anzukurbeln. Deshalb dürfen von jetzt ab keinerlei Beleidigungen oder Beschimpfungen gegen das holländische Volk, die flämischen Volksteile uns. ausgesprochen werden und nur die holländische Regierung bezw. die plutokratischen, herrschenden Kreise, wie Juden usw. angegriffen werden. Die Spaltung zwischen Regierung und Volk soll langsam aber immer stärker in Holland und in den flämischen Teilen Belgiens durch eine wohlgewählte Sprache intensiviert werden.

6. Diary (in a revised form) of Major Engel, *Heeresadjutant* of Hitler, 23–7 May 1940, (see *Heeresadjutant bei Hitler, 1938–1943. Aufzeichnungen des Majors Engel, hrsg. von H. Von Kotze* [Stuttgart 1974], pp. 80–1)

23.5.1940: Telefonat F. [Führer] mit Göring. Feldmarschall der Auffassung, daß grosse Aufgabe der Luftwaffe bevorstehe: Vernichtung der Briten in Nordfrankreich. Heer brauche nur besetzen. Wir wütend, F. begeistert.

25.5.1940: F. hat stundenlang Besuch von Feldmarschall. Berichtet auf Spaziergang von Auffassung G. über endgültige Vernichtung Gegners durch Luftwaffe. Eindruck: G. hat mit Erfolg intensiv gegen Heer gehetzt. F. betont immer wieder weltanschauliche Zuverläßigkeit der Luftwaffe im Gegensatz zum Heer.

26.5.1940: Lagevortrag Jeschonnek. Bestätigung, daß Luftwaffe Vernichtung der Engländer schafft.

27.5.1940: Feldmarschall landet mit Storch. Aussprache mit F. in Lagebaracke. Göring meldet Erfolge im Hafen Dünkirchen, sagt wörtlich: Nur Fischkutter kommen herüber; hoffentlich können die Tommys gut schwimmen.

7. Instructions of the *Oberkommando des Heeres* (*OKH*) to Army Group B, General von Bock, 23 May 1940, 1 a.m. (see *Dokumente zum Westfeldzug 1940*, hrsg. von H.-A. Jacobsen [Göttingen 1960], pp. 116–17).

... Die grossen Städte des Vlamenlandes, auf deren Erhaltung die politische Führung Wert legt, sind möglichst zu schonen.

... Heeresgruppe A schwenkt mit schnellen Kräften nach Erreichen der Linie Béthune–St Omer–Calais ein gegen die Linie Armentières–Ypern –Ostende.

8. *Kriegstagebuch* of Army Group A, General von Rundstedt, entry for 23 May 1940, (see *Dokumente zum Westfeldzug*, hrsg. von H.-A. Jacobsen, pp. 70–1).

Eine neue Weisung des OKH gibt Anordnungen für die Vollendung der Einschließung des Feindes in Nordbelgien und – frankreich. Der Heeresgruppe A fällt dabei die Aufgabe zu … mit den schnellen Kräften über die Linie Béthune–St Omer–Calais gegen die Linie Armentières–Ypern–Ostende einzuschwenken.

9. Kriegstagebuch of Army Group A, entry for 24 May 1940, (see L.F. Ellis, *The War in France and Flanders, 1939–1940* [London 1953], p. 383, and *Dokumente zum Westfeldzug*, hrsg. von H.-A. Jacobsen, pp. 74–5).

Um 11.30 Uhr trifft der Führer ein und läßt sich durch OB der Heeresgruppe über die Lage unterrichten. Der Auffassung, daß ostw. Arras von der *Infanterie* angegriffen werden müsse, die *schnellen Truppen* dagegen an der erreichten Linie Lens–Béthune–Aire– St Omer–Gravelines angehalten werden können, um den von Heeresgruppe B gedrängten Feind 'aufzufangen' stimmt er voll und ganz zu. Er unterstreicht sie durch die *Betonung*, daß es überhaupt notwendig sei, die Panzerkräfte für die kommenden Operationen zu schonen, und daß eine weitere Einengung des Einschließungsraumes nur eine höchst unerwünschte Einschränkung der Tätigkeit der Luftwaffe zur Folge haben würde.

In diesem Sinne wird um *12.45 Uhr* an AOK 4 befohlen.

10. Order given by telephone by Army Group A to Commander of the 4th Army, 24 May 1940, 12.31 p.m. (*Dokumente zum Westfeldzug*, hrsg. von H.-A. Jacobsen, p. 120)

(Fernmündlich von HGr. Kdo A an AOK–4. Ab: 12.31 Uhr)

Auf Befehl des Führers ist der Angriff ostwärts Arras mit VIII. und II. AK im Zusammenwirken mit linkem Flügel H.Gr. B nach Nordwesten fortzusetzen. Dagegen ist nordwestlich Arras die allgemeine Linie Lens–Béthune–Aire– St Omer–Gravelines (*Kanallinie*) *nicht zu überschreiten*. Es kommt auf dem Westflügel vielmehr darauf an, alle beweglichen Kräfte aufzuschließen und den Feind an der genannten günstigen Abwehrlinie anrennen zu lassen.

11. Hitler's Directive (Weisung) of 24 May 1940, (*Hitlers Weisungen für die Kriegführung, 1939–1945*, hrsg. v. W. Hubatsch [Frankfurt 1962], pp. 53–5)

Der Führer und Oberste Befehlshaber   Hauptquartier, d. 24.5.40.
der Wehrmacht
OKW/Abt. L. Nr. 33028/40 g. K. Chefs.
Geheime Kommandosache
Chef Sache   7 Ausfertigungen
Nur durch Offizier   3 Ausfertigung
                    Weisung Nr 13
1. Nächstes Ziel der Operationen ist die Vernichtung der im Artois und in Flandern eingeschlossenen franz. – engl. – belg. Kräfte durch konzentrischen

Angriff unseres Nordflügels sowie die rasche Besitznahme und Sicherung der dortigen Kanalküste.

Aufgabe der Luftwaffe ist es hierbei, jeden Feindwiderstand der eingeschlossenen Teile zu brechen, das Entkommen englischer Kräfte über den Kanal zu verhindern und die Südflanke der Heeresgruppe A zu sichern. Der Kampf gegen die feindliche Luftwaffe ist bei jeder günstigen Gelegenheit fortzusetzen.

3. Aufgabe der Luftwaffe

a) Unabhängig von den Operationen in Frankreich wird der Luftwaffe – sobald ausreichende Kräfte zur Verfügung stehen – die Kampfführung gegen das englische Mutterland in vollem Umfang freigegeben. Sie ist mit einem vernichtenden Vergeltungsangriff für die englischen Angriffe gegen das Ruhrgebiet einzuleiten.

Angriffsziele bestimmt der Ob. d. L. nach den in der Weisung Nr 9 gegebenen Richtlinien und den hierzu von OKW noch zu erlassenden Ergänzungen. Zeitpunkt und beabsichtigte Kampfführung sind mir zu melden.

Der Kampf gegen das englische Mutterland ist auch nach Beginn der Heeresoperationen weiterzuführen.

gez. Adolf Hitler

12. Diary of General Jodl, Chief of the Armed Forces Operations Staff (*Wehrmachtführungsamt* of the *OKW*), entry for 24 May 1940, (*Procès des grands criminels de guerre devant le Tribunal Militaire International, Nuremberg 14 novembre 1945–1er octobre 1946*, vol. 28 [Nuremberg 1948], pp. 433–4)

24.5: Führer fliegt mit mir und Schmundt z(ur) H.Gr. A nach Charleville

...

Es ergeht daher am Abend neuer Befehl:

a) nicht über Linie Sandez [*sic*]–St Omer–Gravelinc [*sic*] nach Osten vorzugehen ...

Führer unterschreibt Weisung Nr 13 über Weiterführung der Operationen

...

Sehr günstige Meldung d[es] Ob.d.L. und Ob.d.H. über die Lage. Es wird offensichtlich, daß der eingeschlossene Feind zu keiner geschlossenen Aktion mehr fähig ist.

13. Diary of General Halder, Chief of the Army General Staff, entry for 24 May 1940, (*Generaloberst Halder, Kriegstagebuch*, vol. i, hrsg. v. H.-A Jacobsen [Stuttgart 1962], pp. 317–19).

Die Lage entwickelt sich weiter durchaus zufriedenstellend, wenn auch das Herankommen von Infanterie-Verbänden in Richtung Arras seine Zeit dauert. Da südlich der Somme zur Zeit keine Gefahr droht, halte ich das nicht für bedrohlich. Die Widerstandskraft des Feindes ist nicht mehr sehr hoch einzuschätzen, abgesehen von örtlichen Kämpfen. Die Dinge werden also ihren Weg gehen; wir müssen nur Geduld haben, sie reifen zu lassen.

15.30 Uhr: Führer ab heute morgen bei Rundstedt.

20.20 Uhr: Befehl, welcher den gestrigen Befehl aufhebt und Einkreisung im Raum Dünkirchen–Estaires–Lille–Roubaix–Ostende anordnet. Der schnelle linke Flügel, der keinen Feind vor sich hat, wird dabei auf ausdrücklichen Wunsch des Führers angehalten! In dem genannten Raum soll die Luftwaffe das Schicksal der eingekesselten Armee vollenden !

14. General (later Field Marshal) Alan Brooke, Commander of II Corps, 'Operations of II Corps during retreat from Louvain to Dunkirk', enclosed with a letter to Churchill of 31 October 1946 (PRO, CAB 106/252).

On the 25th we captured a German Staff car just north of the Lys. The driver of the car was shot, but the Staff officer escaped leaving a large leather wallet behind. I collected this wallet at 3rd Division HQ and brought it to GHQ (near Armentières) where the Intelligence Branch rapidly translated the most important documents. These contained the German plans for an attack by a Corps on to Ypres, a holding attack on the frontier defences East of Lille, and another thrust South of this. This information was of course invaluable.

We had at that time nothing on the Ypres–Comines Canal except one brigade I had obtained from GHQ the previous day. I knew for certain that the Belgians would offer no resistance, consequently a German thrust through Ypres, unless stopped, must inevitably cut off the BEF from the sea.

I have since discovered that the German Staff officer was a Lt.-Colonel (later Lt.-General) Kinzel, who was von Brauschitsch's liaison officer with the Sixth German Army.

15. Diary of General Halder, entry for 25 May 1940, (*Generaloberst Halder, op. cit.*, p. 319).

Der Tag beginnt wieder mit unerfreulichen Auseinandersetzungen zwischen Brauchitsch und dem Führer über die Weiterführung der Einkreisungsschlacht. Ich hatte die Schlacht so angelegt, daß die frontal gegenüber einem sich planmäßig absetzenden Feind zu schwerem Angriff antretende H.Gr. B den Feind lediglich binden, die H.Gr. A., die einen geschlagenen Feind trifft und auf den Rücken des Feindes losgeht, die Entscheidung bringen sollte. Das Mittel dazu waren die schnellen Truppen. Nun bildet sich die politische Führung ein, die letzte Entscheidungsschlacht nicht in das Gebiet der Flamen legen zu wollen, sondern nach Nordfrankreich. Um dieses politische Ziel zu bemänteln, wird erklärt, das flanderische Gelände sei mit seinem vielen Wasser pp. für Panzer ungeeignet. Die Panzer und die anderen schnellen Truppen müßten also nach Erreichen der Linie St Omer–Bethune angehalten werden.

Es tritt also eine völlige Verkehrung ein. Ich wollte H.Gr. A zum Hammer, H.Gr. B zum Amboß machen. Nun macht man H.Gr. B zum Hammer, H.Gr. A zum Amboß. Da H.Gr. B eine festgefügte Front sich gegenüber hat, wird das nur sehr viel Blut kosten und sehr lange dauern. Denn die Luftwaffe, auf die man hoffte, ist vom Wetter abhängig.

Es entsteht aus dieser Verschiedenheit der Auffassung ein Geziehe hin und her, das mehr Nerven verbraucht als die ganze Führungsaufgabe. Wir werden die Schlacht trotzdem gewinnen.

16. Pamphlets distributed to Belgian troops
    a) either directly by the Germans or indirectly through members of the VNV *Militaire Organisatie* (from: Freiburg-in-Breisgau, Bundesarchiv, Militärarchiv)
    One example:
    VLAMINGEN KAMERADEN!!
    We beleven een geweldige omwenteling in het oude Europa!! Twee wereldbeschouwingen botsen tegen elkaar ... Aan de eene zijde de ARBEID op den troon – Aan de andere zijde het GELD – Frankrijk en zijn vazaalstaat Belgie hebben reeds lang in naam van Vlaanderen partij gekozen! Men zegde – schreef – riep-NEUTRALITEIT!! Doch in gezamenlijk overleg bouwde men EENZIJDIGE versterkingen tegen Duitschland ... Tegen het Jonge Nieuwe Duitschland, Kameraden, dat zooveel recht heeft op zijn plaatsje onder de zon als Engeland of Frankrijk! Tegen een Germaansch broedervolk, Kameraden, dat zich NOOIT VERGREEP OF ZAL VERGRIJPEN aan ons VOLKSZIJN aan onze TAAL en ZEDEN.
    KAMERADEN!!! Thans wordt geschiedenis geschreven ... Zal Vlaanderen weer gelooven aan ijdele woorden en beloften? Zal het zich weer lijdzaam op sleeptouw laten nemen door de eeuwenoude erfvijand Frankrijk –
    THANS MOET GEKOZEN WORDEN!!!
    Met al wat Waalsch en valsch is tegen het broedervolk waaraan we met zoovele vezels van ons volkszijn verbonden zijn?
    NEEN, Kameraden! DAT NIET WEER!! Wie nog een greintje Vlaamsche fierheid in zich voelt rukke zich voor goed los van het volksvreemde en volksvijandige FRANCO-BELGIE! Eens hebben Belgische staatsministers in Frankrijk beloofd ... als het et zal op aan komen
    MARSJEERD VLAANDEREN voor FRANKRIJK ...
    nu komt het er op aan ...
    BEWIJST HET TEGENDEEL!
    b) directly by the Germans, before the capitulation of the Belgian army of 28 May 1940, (see K. Kirchner, *Flugblätter. Psychologische Kriegsführung im Zweiten Weltkrieg in Europa* [München 1974], no. 20).
    VLAAMSCHE SOLDATEN!
    Elke uur van nutteloozen weerstand kost vreeselijke verliezen.
    Vecht ge voor de Franschen, die nooit uwe vrienden waren?
    Neen!
    Voor de Engelschen, die op hun laffe vlucht naar Kales uwe steden en dorpen ultplunderden?
    Neen!

Wilt ge zooals in den wereldoorlog in de voorste rij voor anderen verbloeden?

Neen!

Houdt op met schieten!

Geeft u over!

Redt u voor uw land en uw gezin. Ge zult goed behandeld worden! Dan redt ge ook Vlaanderen van de verschrikkingen van dea oorlog.

ALLES VOOR VLAANDEREN!

17. Diary of General Halder, entry for 28 May 1940, (*Generaloberst Halder, Kriegstagebuch*, i, p. 323).

Im Laufe des Vormittags ist OB beim Führer und kommt von diesem 13.00 Uhr mit folgenden Festsellungen zurück: (...)

d) Liquidation Belgien:

1. Akt: Entwaffnen, verpflegen, Pulk (= Masse der Kriegsgefangenen auflockern in Richtung Antwerpen–Brüssel ...

2. Akt: Flamen und Wallonen (i.e.: unter Kriegsgefangenen) trennen. Flamen entlassen. Wallonen wahrscheinlich abtransportieren, Arbeitskräfte zu Hause.

18. Weekly Political Intelligence Summary Nos 34, 29 May 1940 (Great Britain, Foreign Office, Weekly Political Intelligence Summaries, vol. i, October 1939-June 1940 [London 1983]).

German attempts to sow dissension between Walloons and Flemings appear to follow the lines which became familiar between 1914 and 1918. Recently German newspapers are reported from Switzerland as publishing maps showing the division of the country into Walloon and Fleming [sic] areas. The argument is once more advanced that the combination is unnatural, and an appeal is made to Flemish nationalist aspirations by suggestions that sections of North Eastern France, including Lille, are historically Flemish in origin and would therefore fall naturally within a new Flemish State to be organised under German protection.

19. Instructions of Colonel Böhme, of the *OKW*, to the military Government in Brussels, 4 June 1940, (see A. De Jonghe, *Hitler en het politieke lot van België, 1940–1944*, vol. i [Antwerp 1972], pp. 116, 380–1 n. 14)

Oberstlt. Böhme berichtet folgendes: ... Der Führer beabsichtigt, für die flämischen Gebiete später einen Reichskommissar einzusetzen; vorläufig jedoch nicht. Die Wallonischen Teile, sowie Frankreich, werden nach der Absicht des Führers unter einer Militärverwaltung bleiben.

20. Diary (in a revised form) of General Van Overstraeten, military adviser of King Leopold III, entry for 4 June 1940, (Général Van Overstraeten, *Léopold III Prisonnier* [Bruxelles 1986], pp. 19–20).

A 18 heures, le docteur Gebhardt demande à me parler. Il désire s'exprimer en allemand, par souci de précision…:

'Nous sommes avant tout nationaux-socialistes … Il importe de voir toutes choses de haut et en profondeur. Entre autres, la question des prisonniers est très complexe. Il s'y mêle en effet, en essence première, le problème raciste. Nous entendons créer un immense espace vital, homogène de race et économiquement viable … Le problème des prisonniers est complexe; car il sera résolu dans le sens racique. Les soldats flamands seront libérés; les soldats wallons resteront en captivité. Il faut s'y résigner. S'il n'y avait eu que des Flamands dans l'armée, nous serions entrés en Belgique sans tirer un coup de fusil. La Flandre avait perdu le souvenir de la guerre de 1914–1918. Au contraire, les Wallons l'avaient gardé vivace … Nous parlons la même langue que les Flamands, à peu de chose près. Nous ne voulons pas de gens hétéroraciques dans notre association.'

J'ai écouté avec une peine indicible cette diatribe … J'enrage sous l'oeil en vrille de cet impudent morticole qui annonce avec joie le dépècement de la Belgique. Me maîtrisant par devoir, je proteste contre l'insinuation que la majorité des Flamands aurait accepté l'invasion sans opposer de résistance.

21. Diary of Van Overstraeten, entry for 8 June 1940, (Van Overstraeten, ibid., p. 25).

Kiewitz me lit une instruction définitive du général von Falkenhausen concernant le sort de l'armée:

'Le Führer a décidé, le 5 juin, que tous les prisonniers de guerre issus des provinces de la Flandre occidentale, de la Flandre orientale, d'Anvers, du Limbourg, du Brabant (ville de Bruxelles et arrondissement de Nivelles non compris) seront immédiatement relâchés. Les prisonniers originaires des autres provinces, c'est-à-dire des territoires wallons, seront transportés en Allemagne. Toutefois, les prisonniers de ces territoires pouvant établir leur origine flamande, ou exerçant une profession d'utilité publique, seront libérés immédiatement'.

22. Minute of a meeting held on 19 June 1940 at the headquarters of Field Marshal Göring, (see J. Gumkoeski and K. Leszczynski, *L'Occupation hitlérienne en Pologne* [Warsaw 1961], photograph between pages 32 and 33, and *Procès des grands criminels de guerre*, vol. xxvii, p. 30).

Die Rückführung der in der besetzten Gebieten vorgefundenen Rohstoffe ist mit äusserstem Nachdruck zu betreiben … Auf landwirtschaftlichem Gebiet steht im Vordergrund die Rückführung von Vieh aus Nordfrankreich und Belgien. Eine Fortnahme von Vieh bei den Vlamen soll unterbleiben.

23. General Keitel, Chief of the *Oberkommando der Wehrmacht* (*OKW*), to General von Brauchitsch, Commander-in-Chief of the Army, 14 July 1940, (*Akten zur Deutschen Auswärtigen Politik, 1918–1945*, series D, 1937–45, vol. x/3 [Frankfurt 1963], no. 167, p. 174).

Chef OKW   14. Juli 1940
Nr: 1330/40 g. K.   Chefsache
Geheime Kommandosache   Nur durch Offizier
Der Führer hat hinsichtlich der Zukunft des belgischen Staates noch keine
endgültige Entschließung getroffen. Er wünscht einstweilen jede mögliche
Förderung der Flamen einschl[ießlich] Rückführung der flämischen Kriegsge-
fangenen in ihre Heimat. Den Wallonen sind keinerlei Vergünstigungen zu
gewähren.

24   Minute of an interview between Hitler and Princess Marie-José of Italy, sister
of King Leopold III, 17 October 1940, *(Staatsmänner und Diplomaten bei Hitler,*
1939–1941, hrsg. v. A. Hillgruber [Frankfurt 1967], p. 256).

Der Führer erwiderte:

In der Gefangenenfrage handele Deutschland lediglich aus der Sorge eines
Staates heraus, dem ein Krieg auf Leben und Tod aufgezwungen worden sei.
Daher habe es bisher nur solche Gefangene freigelassen, von denen es
annehme, daß sie ihrer Mentalität und Einstellung nach nichts gegen Deutschland unternehmen würden. Auf diese Weise seien vor allen Dingen flämische
Gefangene freigelassen worden. Deutschland habe kein Interesse daran, durch
allgemeine Freilassung der Gefangenen Belgien möglicherweise eine Injektion
chauvinistischer Feindseligkeit zu machen.

25.   Speech of Staf De Clercq, leader of the *Vlaamsch Nationaal Verbond (VNV)* on 10
November 1940 in Brussels, *(Volk en Staat,* 13 November 1940).

Tientallen van malen heb ik gezegd op onze grote volksvergaderingen of
geschreven in onze bladen: 'Het Vlaamsch Nationaal Verbond zal er voor
zorgen dat de Vlaming geen bloed zou vergieten voor vreemde belangen'. In de
konjunktuur waarin we leefden beteekende zulks: 'Geen Vlaamsch bloed voor
Fransche of Engelsche belangen'. Die belofte hebben wij gehouden. Als er in
de maand Mei geen duizenden Vlaamsche en duizenden Duitsche soldaten
meer zijn gevallen, dan kan het Vlaamsch Nationaal Verbond zulks gerust of
zijn rekening schrijven! Dat kunnen wij ook bewijzen. Doch de tijd is nog niet
gekomen om die schitterende bladzijde van Vlaanderen's geschiedenis open-
baar te maken.

26.   Minute of the interview of 19 November 1940 between Hitler and King
Leopold III *(Staatsmänner und Diplomaten,* p. 342).

Der Führer erwiderte daß Deutschland die Arbeit der Kriegsgefangenen
dringend benötige … Man habe jedoch in Aussicht genommen, einige Belgische Kriegsgefangene zu entlassen, und zwar diejenigen, von denen bekannt sei,
daß sie bereits früher Deutschland gegenüber eine wohlwollende Haltung ein-
genommen hätten. Es würde sich dabei vor allem um Flamen handeln.

EMBATTLED FEMININITY: CANADIAN WOMANHOOD
AND THE SECOND WORLD WAR

Ruth Roach Pierson

When the Canadian government declared war against Germany on 10 September 1939, Canada was still recovering from the Great Depression.¹ Initially the thousands of unemployed met the expanding 'manpower' needs of war production and military recruitment. Within one year, however, National Defence Headquarters had begun to think of using female labour to meet a shortage of well-qualified male military personnel for specific jobs, such as those of clerk and cook. In the course of July 1941 to July 1942, official women's services were created within all three branches of the Canadian armed forces: in July 1941, the Canadian Women's Auxiliary Air Force, later changed to the Women's Division of the Royal Canadian Air Force; one month later, the Canadian Women's Army Corps (CWAC); and, in July 1942, the Women's Royal Canadian Naval Service. Thus, for the first time in the history of Canada, women other than nursing sisters were admitted into Canada's military. By war's end, nearly 50,000 women had served in the Women's Services of the Canadian armed forces. In early 1942, the pre-war slack in the civilian labour market was completely taken up and the Department of Labour together with the newly created National Selective Service turned to 'womanpower' (a term of wartime coinage)² to relieve the 'manpower' shortages beginning to plague war industry. At the height of the war effort, between one million and 1,200,000 women were working full time in Canada's paid labour force, fully twice the number that had been there in 1939. Approximately 260,000 of these female workers were directly involved in war production, and some proportion of them were holding non-traditional jobs. This paper examines the shifts in gender ideology accompanying these wartime developments as they can be traced in the public discourse and in the pronouncements of the National Selective Service and of the Departments of Labour, National Defence, Munitions and Supply, and National War Services, and of media supportive of Canada's war effort.

'They're Still Women After All' was the title given, with an almost audible sigh of relief, by Canadian foreign correspondent Lionel Shapiro to a piece he wrote for a major Canadian magazine in September 1942. In a jocular vein, the article expressed one man's fear, aroused by the wartime sight of large numbers of British women stepping into formerly male jobs: the fear that women might cease to be women, that is 'feminine individuals', synonymous terms in Shapiro's vocabulary. Similar fears plagued many other Canadians, male and female, as they viewed women entering the munitions plants and, what in the eyes of some seemed even more disruptive of the gender *status quo*, joining the armed forces.[3]

As the primary purpose of the military services is the provision of the armed might of the state, their male exclusivity had been in keeping with a deeply rooted division of labour by sex that relegated women to nurture, men to combat, women to the creation and preservation of life, men, when necessary, to its destruction. Closely connected to the sexual division between arms bearers and non-arms bearers was a gendered dichotomy of attributes that identified as masculine the military traits of hardness, toughness, action and brute force and as feminine the non-military traits of softness, fragility, passivity and gentleness. Hence, the very entrance of women into the Army, Navy, and Air Force sharply challenged conventions respecting women's nature and place in Canadian society.

While it took impending manpower shortages to induce the Department of National Defence to admit women into the armed services, there had existed across Canada since 1938–9 a small but determined paramilitary movement of women volunteers eager for military service. And while the members of the women's volunteer corps keen to become part of Canada's official forces wanted to end women's exclusion from the military, for the men in charge of Canada's military, efficient prosecution of the war was the reason for putting women in uniform and under service discipline. They had no desire to tamper with existing sex/gender relations by altering the sexual division of labour or the male over female hierarchy of authority. In the case of the somewhat less dramatic changes brought about by women's admission into non-traditional trades in war industry, there is evidence that some women would have liked to consolidate the sex/gender boundary shifts occasioned by the war for continuation into the post-war world. But the widening of the range of jobs to which women could gain access during the war was driven, not by a movement for women's right to work, but rather by the exigencies of a wartime economy.

There was a tension inherent in the admission of women into the armed services: the tension between the Canadian State's wartime need for female labour within those pre-eminently masculine institutions and Canadian society's longer term commitment to a masculine-feminine division of traits as well as separation of tasks. This tension was also apparent, albeit to a lesser degree, in the entrance of women into non-traditional trades in war industry, and in the eventual open recruitment of married women into paid employment.

How did the official discourses of the war departments of government and the mainstream media construe women's entrance into the armed forces as well as into non-traditional trades in war industry? Appropriating a feminist discourse largely buried during the Depression years, one construction was that the war was opening up for women a world of opportunity unrestricted by sexual inequalities. The ceremonial launching of a ship that women workers had helped to build 'from the first bolts and staves to the final slap of paint and piece of polished brass' prompted journalist Lotta Dempsey to hail the event as symbolising the launching of 'the great and final stage of the movement of women into industry ... on a complete equality with men'.[4] Army recruiters, too, spoke of the war's having 'finally brought about complete emancipation of women'.[5] Evidence for this was to be found in the very admission of women to the forces and to an ever increasing number of trades within the forces. With the punning title '"Jill" Canuck Has Become CWAC of All Trades', a 1944 magazine article celebrated the expanding number of trades open to members of the Canadian Women's Army Corps and singled out in illustration the first 'qualified girl armourer in the CWAC' and the first CWAC to operate 'a Telecord recording machine'.[6]

The mass advertising media also participated in the celebration of women's wartime freedom. For example, an advertisement placed by General Motors of Canada in a number of mainstream Canadian magazines proclaimed the end of women's confinement to the domestic sphere. Running for five pages it was set up so that a reader would encounter 'Woman's place ...' on a right-hand page and then turn to find not 'is in the home' but 'is Everywhere ...', followed on the fourth and fifth pages by the modifying prepositional phrase 'with Victory as Their Business' and a brief description of the many fields in which women were serving the nation at war.[7]

At the same time, war industry and the federal government through its war departments and the National Selective Service had to be sensitive to what were perceived to be the dominant values of the larger civilian society, especially given the war effort's dependence on women volunteers, not conscripts. To inform itself of public attitudes, the federal government made use of opinion surveys and to influence public opinion it made use of the media. At the same time, as pillars of the established order, those staffing the various state agencies and writing for the mainstream media unexaminedly embodied and uncritically gave voice to dominant notions of women's nature and capabilities. The wartime federal government was caught in something of a dilemma: by admitting women to the Services and into non-traditional trades, it had violated the convention that the armed forces and the shop floors of heavy industry were no places for women, and by openly drawing on married women's paid labour, it had violated the dictum, consecrated as 'common sense' during the Depression, that a married woman's place was in the home. Government policy sought to minimise or disguise these violations by advertising its conformity with as many received notions of proper womanhood as possible. Moreover, while Canada's mobilisation of women for the war effort challenged the social ideal of the woman dedicated to home and family, at the same time imagery of women's dedication to home and family was deployed to symbolise what Canada was fighting for and to present, in contrast to the war's cruel disruption and destruction of human lives, a picture of stability realisable, if not in the present, then in the post-war future. Thus, to those in charge of the war effort and to many in the media, efficient prosecution of the war required both disruption of conventional notions of women's place and containment of that disruption. And thus, with respect to women's participation in the armed forces and in non-traditional trades and married women's departure from the home for the paid work-force, alongside the talk of emancipation, equality and the overcoming of tradition, recruitment propaganda and wartime advertising also sought to set limits to the degree of change required.

The concept of 'war emergency' was repeatedly invoked to justify most of the wartime changes widening women's job opportunities and accommodating their wartime roles. For instance, the 1942 income tax concession to married women was rationalised as a war measure, 'justified only by the extreme state of emergency which then existed'.[8] Moreover, federal agencies viewed the Dominion-Provincial Wartime Day Nurseries Agreement, negotiated with Ontario and Quebec, as a war emergency measure designed 'to secure the labour of women with young children' for 'war industry'[9] and to stay in effect only 'for the duration'. Indeed, the deployment of the vocabulary of 'war emergency' and 'for the duration' served as one of the most effective means of temporally containing women's expanded sphere of operations.

Similarly, a major genre of wartime advertising imaged a rapid return to normality once the war was over. For example, Westinghouse of Canada provided a graphic statement of the expectation that at war's end women would return to motherhood and child-rearing as their principal life's work. A large two-page advertisement headed 'These Are Tomorrow's Yesterdays' showed the pride the future son of the female war-worker would take in his mother's war service. Looking through a book on *Women at War* in the school library in 1955, the boy was surprised by a picture of his Mom, taken in a war plant in 1943. Overcome with pride, the child has sat down to tell his Mom in writing how proud of her he is. The advertisement carried not only a tribute to Canadian women's role in Canada's struggle, but also the projection that in ten or twelve years' time the women would be back where they belonged, taking care of the home, rearing Canada's future generations, and buying Westinghouse appliances.[10]

Recruitment propaganda, promotional newspaper stories and patriotic advertising, then, reveal a double sidedness or, one might say, a contradiction, in the treatment of women's wartime mobilisation. On the one hand was the celebration of the trail blazing and achievement of women in the services and in non-traditional work; and, on the other, the message that joining the forces or participating in war industry entailed no permanent change in women's nature or place in Canadian society.

In 1943 the message of reassurance was intensified with respect to women's recruitment into the armed forces as a need for more women in the military was anticipated but monthly enlistment figures dropped. Faced with this slow-down, the National Campaign Committee granted authorisation to two opinion surveys in the first half of 1943:[11] first, a general survey of public opinion,[12] and second, a study of the opinion of CWAC other ranks.[13]

Both surveys revealed that there was public disapproval of women joining the armed forces. One source of disapproval identified was the fear that the woman who joined the forces did so at the risk of her femininity. In the hopes of dispelling that notion, speeches of recruitment officers, recruitment literature, sponsored advertising and promotional news stories were full of assurances that membership in the women's services was not incompatible with femininity. 'Our women in the Canadian Armed Forces are nothing if not thoroughly feminine in manner and appearance', recruiting officers instructed National Selective Service personnel being trained to assist with the recruitment of servicewomen.[14] Clearly, while the armed forces could persuade a male potential recruit that military service would make a man of him, they felt they had to do the reverse with a female prospective volunteer, that is, convince her that military life would not make her less of a woman.

Catering to women's assumed concern with their appearance was one strategy adopted by military authorities. Competition developed among the three official Women's Services as to which could claim the uniform with the best design. Having engaged the designing talents of one of Toronto's most exclusive women's clothing stores, the Canadian Women's Army Corps boasted in its 1942 recruiting brochure: 'C.W.A.C. uniforms have been acknowledged by leading dress designers to be the smartest in the world.'[15]

Women's pride in uniform, though, had to be cultivated along rather different lines from men's pride in uniform. A long association linked male with military uniform and military uniform with the masculine traits of forcefulness, toughness, and, in the case of an officer's uniform, commanding authority. When a man donned a uniform he stood to see his masculinity enhanced. Hence the popular notion that uniforms (on men implied) turned women's heads. Literary evidence exists that where the war effort was strongly supported and there were large concentrations of men in uniform, men of military age not in uniform felt threatened not only by the possible charge of cowardice but by the possible loss of girl-friends to the soldiers, sailors and airmen.[16]

Just the reverse was the fear in the case of servicewomen in uniform (or for that matter, of women war-workers in overalls and bandanas). Here, it was believed, reassurance was needed that such garb did not diminish femininity. A magazine article appearing in 1943 entitled 'They're Still Feminine!' sought to convince readers that 'khaki, Air Force and Navy blue, or well-worn denim and slacks' had no lasting effect 'on the softer side of Womanhood'.

> Clothes don't make the man
> and uniforms and overalls
> don't seem to be unmaking
> the female of the species

was set out in bold print. Accompanying the article was a cartoon showing a group of women war-workers and servicewomen congratulating a member of the Women's Division of the RCAF, who blushingly holds out a diamond engagement ring for the others' admiration.[17]

Although eventually the overalls and bandana of the women war-workers became symbols of service, in the early stages of women's recruitment for war industry government officials were concerned that disapproval of the gender violation entailed by women wearing trousers would keep women from applying for essential jobs. In March 1942, the Department of Munitions and Supply for Canada considered it necessary to advertise the plea 'PLEASE DON'T STARE AT MY PANTS' as issuing from a woman war-worker and to illustrate it with the following picture: a middle-aged matron in hat, gloves and furs, has stopped to gaze disdainfully at the young woman in slacks, while a man in a white collar and overcoat looks down with a condescending smirk on his face. The fine print had the object of all the attention replying in this diffident fashion: 'Would you like to know why I wear trousers like the men when I go about the streets? Because I'm doing a man's job for my country's sake.'[18]

A mere year and a half later, however, attitudes had changed enough that *National Home Monthly* could put on the cover of its September 1943 issue a humourous drawing that reversed the relations depicted in the earlier poster. Now it is the female war-worker, confidently swinging her lunch pail while striding along in her slacks, who wheels around to take a second rather scornful look at a passing kilt-clad member of a Highland regiment. The growing popularisation of pants for women can be traced to the wearing apparel required of women working on the shop floors of armament, munitions and aircraft assemblage plants.

The authorisation of slacks for female military personnel, however, remained slow and restricted. Only in 1944 did the Navy's traditional bell bottoms begin to be issued for wear to members of the Women's Royal Canadian Naval Service, and then only to a select number of Wrens on coastal duty. To announce the new departure, a specially-taken photograph was released to newspapers across Canada in June 1944. When it appeared in the *Toronto Telegram*, it carried the following caption: 'No wonder this husky sailor seems slightly puzzled. He has just spied a WREN wearing, of all things, bell-bottomed trousers!'[19]

Similarly, in the Army, the wearing of trousers by women had to be justified by working conditions or by the task to be performed, but even so widespread extension was contested. For instance, trousers were authorised early on as part of the winter (but not summer) uniform for CWAC drivers of trucks and jeeps, but not of staff cars.[20] Then in the summer of 1944 the Master-General of the Ordnance authorised 'Trousers ... for summer wear by CWAC drivers' but 'of Trucks and Jeeps *only* ', and went on to specify in no uncertain terms that CWAC drivers of staff cars would have to continue to wear skirts, both summer and winter.[21] The serious attention given in these regulations as to when servicewomen on duty could or could not wear trousers provides a good example of the wartime politics of dress and of the symbolic power invested in attire as a regulator of gender relations.

One of the least examined and most unshakeable notions of the time about women was that subordination and subservience to men were inherently female characteristics that dictated women's role and place in society. This assumption converged with the actual position of most servicewomen in relation to servicemen, for in jobs, pay and benefits and place in the command structure of the service, the servicewomen were in general subordinate. As reflected in mottoes and enlistment slogans, the very function of the women's services was to subserve the primary purpose of the armed forces: the provision of an armed and fighting force. Having been excluded/exempted from combat, the women of the Canadian Armed Forces could adopt as their motto: 'We Serve That Men May Fight.'[22] That general motto had been adapted from the airwomen's 'We Serve That Men May Fly.'[23] Even more expressive of the secondary status and supportive role assigned to Army women was a slogan used on enlistment advertisements: 'The CWAC Girls — The Girls behind the Boys behind the Guns'.[24]

In preserving the male monopoly on combat service, the Department of National Defence was acting in harmony with social convention and conviction. The prevailing view was that men were by nature suited to dangerous, life-risking jobs while women were naturally adapted to monotony and behind-the-scenes support work, as reflected in these remarks of one air officer on the suitability of airwomen for the trade of parachute rigger:

Take parachute packing. To a man it's a dull, routine job. He doesn't want to pack parachutes. He wants to be up there with one strapped to his back. But to a woman it's an exciting job. She can imagine that someday a flier's life will be saved because she packed that parachute well. Maybe it will be her own husband's life or her boy-friend's. That makes parachute packing pretty exciting for her and she does a much more efficient and speedy job than an unhappy airman would.[25]

Deeply entrenched as the assumption was that the female was the second sex, there still surfaced from time to time the fear that women who were serving with the armed forces would lose their deference toward men and become 'bossy'. Mainly it was the prospect of female officers that seems to have aroused this fear. When Lionel Shapiro covered Major Alice Sorby's arrival in London, England, he approved of the fact that she was 'pretty' and had 'graciously submitted' to the press conference. But at the perception that she was also 'a woman full of the barbed-wire quality of a colonel in the Indian service', Shapiro confessed, 'my man's world was beginning to totter before my eyes'. But a question of his that 'penetrated her military facade and revealed her in all her feminine vulnerability' saved the day for Shapiro. 'Major Sorby', he asked, 'your husband is a lieutenant and you are a major. I assume he will have to salute you when you meet. Is that not so?' After a moment's thought, Major Sorby replied that, yes, she supposed that would be regulations, but as she had not seen her husband for more than a year, she doubted whether they'd 'bother about salutes'. Missing the irony of Major Sorby's retort, Shapiro was jubilant. 'Moment triumphant!' he exclaimed:

My man's world returned in full flower. Major Sorby was really only Mrs Sorby with a King's crown on her epaulettes, and Mr Sorby, though a mere lieutenant, was still master of the Sorby household.[26]

The apprehension that women serving in the forces would lose their femininity was rooted in the fear that the male over female hierarchy of authority might be upset.

Just as importantly the public had to be assured that the work women in the forces were asked to perform was suitable to women. With an ambivalence typical of the promotion of the women's services, speeches or publications that in one sentence applauded women's breakthrough into positions theretofore dominated by men in the next denied that servicewomen were being asked to do anything inimical to their feminine nature. 'If a woman can drive the family car, she can drive a staff car' is an example of the recruitment line that sought to reassure the public that military jobs for female personnel, although performed in new settings and under different conditions, remained essentially 'women's work'.[27] Thus the contradiction between the armed services' need for women in uniform and the ideology of woman's place being in the home (or in a paid job long sex-typed as female) was bridged through the redefinition of certain military jobs as womanly.[28]

In actual fact, the overwhelming majority of uniformed women employed by the armed services were assigned to jobs that had already become female niches in the civilian labour market or could be regarded as extensions of mothering or housework.[29] In propaganda and practice, a 'woman's place' was created within the wartime armed services. Similarly language and image manipulations were performed in order to 'domesticate' women's employment on dangerous jobs in munitions factories. While one woman war-worker appears to 'be preparing a batch of cookies ... actually she is laying out ignition caps for smoke cartridges'.[30] This 'domestication' of women's war industrial work like the 'womanisation' of women's military assignments both legitimated and curtailed, as historian Ruth Milkman has argued, the 'redefinition of the boundaries between "women's work" and "men's work"' during the war and helped facilitate the return to pre-war patterns of occupational segregation by sex during post-war reconversion.[31]

Even before the opinion surveys of 1943 registered fear of the loss of femininity as a source of ambivalence towards women in uniform, an ugly wartime 'whispering campaign' was underway in Canada, alleging that military service for women invited promiscuity. In the dominant discourse on femininity, respectability was contingent on the preservation of at least the appearance of chastity. The fear or suspicion of its loss, therefore, was always hovering on the pheriphery of the debate over the impact of wartime service on Canadian womanhood. The Wartime Information Board's March 1943 report on the 'whispering campaign' provided an historical explanation for the imputation of 'loose' morals to servicewomen, arguing that sexual respectability had been for centuries 'woman's vulnerable point, the traditional focus of attack by those who resent any extension of her prerogatives'.[12]

Gravely concerned that the rumours of servicewomen's licentiousness were having an adverse effect on enlistment and morale, officers at Defence Head-quarters elected to cover up the existing evidence of VD infection or 'illegiti-mate' pregnancy among servicewomen, rather than to publicise their low inci-dence.[13] Instead, alongside the introduction of measures to screen potential recruits more selectively, public relations officers sought to persuade the public that the 'girls' in uniform were well looked after. The public was told, for instance, that officers commanding servicewomen's units operated *in loco parentis*, and service policewomen acted as 'uniformed guardians of the morals and manners of the Women's Army and Air Force'.[14] At the same time adver-tisements proclaiming parental pride in a daughter's joining up were designed to counter the anticipated parental fear for a daughter's sexual safety. 'I'm proud of you, Daughter' was scrawled in large writing across the top of one advertisement for the Canadian Women's Army Corps. Below appeared the drawing of a father with his arm around a sweet-faced and smiling young woman in uniform.[15] Representations of the piety of Canada's servicewomen further served to underscore their propriety, as in an official photograph of two CWACs singing hymns in a church[16] and in a poster of CWACs attend-ing Sunday service even when stationed overseas.

If the official discourse on wartime femininity oscillated between emanci-pated and traditional womanhood, the military discourse on female sexuality was even more fraught with contradictions. The less cumbersome and more comfortable female wearing apparel of the 20th century facilitated women's freer movement, but at the price of an increase in possibilities for sexual objectification. While the one-piece bathing suit of the 1920s had transformed the North American woman 'bather' into a 'swimmer', it also made possible the 'bathing-beauty' contest and its commercial exploitation.[17] If anything, the war accelerated the twentieth-century's sexualisation of women's images in popular culture and mass advertising. On the inside of the lockers and on the nose cones of planes, servicemen 'pinned-up' photos of female movie-stars and models or drawings of imaginary beauties, usually scantily clad in bathing suits or less. The *CWAC News Letter* encouraged Corps members to participate in 'pin-up' contests run by KHAKI, the Army's bulletin for male soldiers and the parent organ of the CWAC publication. According to literary historian Susan Gubar, women's eroticised images were conscripted during the war to 'reinstate the man's sense of his masculininity' and to serve both as compensation, and as object of retaliation, for men's suffering and sacrifice.[38]

In the 1920s the middle-class National Council of Women of Canada had raised its collective voice against 'the exploiting of Sex Appeal' in the mass media.[39] In the heyday of the 'pin-up girl' during the Second World War, military officers in public relations never gave a second thought to using female 'sex appeal' to sell the women's services or, for that matter, military hardware. Public relations officers singled out servicewomen with the beauty queen or movie-star good looks of the day to promote the women's services, as photographs of CWAC personnel arranged in typical 'pin-up' poses demonstrate.[40] Military public relations also used what were then called 'glamour shots' to publicise events in support of the war effort. For example, to draw attention to an exhibit of armaments in Ottawa in May 1944, Army photographers took a series of shots of the exhibited war equipment with which they posed a pretty CWAC corporal. The caption to one photograph in the series read:

The huge spot-light on display with a collection of modern war material in Ottawa, has a valuable use in searching out enemy night raiders, but Cpl. Wilma Williamson, CWAC of Dundas, Ont., has found another use for it ... it provides an excellent reflection for straightening a crooked Service tie.[41]

A lot was conveyed by that photograph and caption, above all the contrast between the serious operational function of the war equipment and the ornamental function and irrepressible narcissism of the CWAC.[42]

War industry also traded in female sex appeal, intermingling the exploitation of sexual attractiveness with the campaign to reassure women that warwork went hand in hand with feminine 'glamour'. Munitions, armament and aircraft plants staged annual 'beauty contests' to raise the morale of their women workers and to parade the company's patriotism before the public.

Persuading the public of their protectiveness of respectable femininity while at the same time exploiting eroticised images of women required something of a balancing act from the publicity for war industry and the military. To dispel the notion that a woman's attractiveness vanished on enrolment in the armed forces, for instance, military authorities were prepared to make concessions in the area of make-up. The 1920s had seen the introduction of artificial cosmetics as commodities to be purchased on the market. Adoption by women of the middle and upper classes had established their use as fashionable, if applied with restraint. Excessive use was associated with less respectable women, especially prostitutes. The regulations for the Canadian Women's Army Corps permitted women in uniform to wear make-up, so long as it was applied lightly and in moderation. Similar restraint was to govern the use of nail varnish: clear or lightly tinted shades were all right but 'highly coloured' polish was not.[43] While on the one hand military authorities felt constrained to regulate against possible 'vulgarity', on the other hand, the personal decision of the first Officer Administering the CWAC, Lieutenant Colonel Joan B. Kennedy, to wear no make-up nor jewellery at all, was regarded as unnecessarily severe.[44] Concern lest the servicewoman acquire the image of a 'painted lady' was counterbalanced by concern lest she appear manly.

Both concerns lay behind the public-spirited decision of the National Council of the YWCA to offer 'beauty culture' classes at their Women's Active Service Club in Ottawa. 'Seated at long tables before individual mirrors', the servicewomen were 'taught the importance of good skin grooming.'[45] 'Rouge should be natural looking' was one piece of advice handed out by the 'well-known beauty consultant' who acted as 'volunteer instructress' of the classes.[46] Open to women of all three services, the classes were photographed by the military to use as further proof that enrolment in the armed forces was not detrimental to genteel feminine beauty.[47]

The 'attractiveness' so central to the prevailing notion of femininity ultimately meant attractiveness to men. Heterosexuality was certainly one of the norms of femininity subscribed to in official publicity about the women's services. While advertisements acclaimed the good wholesome fun women could have together in the services,[48] celebrations of female camaraderie stopped short of any suggestion of lesbian intimacy. Under military law, same-sex relationships could be and were construed as grounds for discharge.[49] Indeed the concern whether life in the armed forces made a woman more or less attractive boiled down to a concern whether enlistment made a woman more or less marriageable.

In Army, Navy and Air Force photo stories on life in the women's services, public relations made a point of including material on dances and dates and boy-friends. The message of an Army photo series on CWACs in Washington, DC, was that members of the Canadian Women's Army Corps stationed outside Canada could find dates in the armed forces of Canada's allies. One photo showed two CWAC privates stepping out with a soldier and a sailor of the United States Forces.[50] Army, Navy and Air Force public relations encouraged news coverage of marriages between servicemen and servicewomen.[51] Announcements of such military weddings became a regular feature of the women's activities columns of Canadian newspapers.[52] Publicising the closeness of attachments between servicemen and servicewomen served to enforce heterosexual norms as well as to downplay reports of the ill-feeling entertained by some servicemen toward the women's services. At the height of the 'whispering campaign', however, the Wartime Information Board viewed even the widely used 'shoulder to shoulder slogan' as ill-advised, since it implied that women were 'mixed up with men', and recommended its withdrawal, a drastic measure as this was the title of the official recruiting song of the Canadian Women's Army Corps.[3]

Caught between a patriotic desire to make an all-out effort to win the war and a conservative unwillingness to change women's relegation to home and family (or to temporary and lowly places in the paid labour force), the Canadian State and Canadian society responded with ambivalence to the admission of women into the armed forces and into non-traditional trades during the Second World War. The visible movement of women out of the home into apparently greater economic and sexual independence touched off fear of an impending breakdown of the sexual division of labour.[4] Humour of various sorts provided an outlet for these fears. In a 1942 magazine article entitled 'Woman Power', for instance, a female journalist jokingly proposed a number of ludicrous possibilities. What if at war's end, she speculated, the thousands of war-working women

refuse to be stripped of the pants and deprived of the pay envelopes? What if they start looking round for some nice little chap who can cook and who'll meet them lovingly at the door with their slippers in hand? What if industry has to reorganize to give these women sabbatical years for having babies?[5]

The fear of the loss of femininity occasioned during the Second World War by the sight of women in uniform or slacks and bandanas betokened a fear of structural changes in the male dominant sex-gender system — a fear of women slamming the door on domestic dependence, assailing the segregation of jobs by sex, and asserting sexual independence. During the latter part of the war, those fears were appeased by the measures of containment discussed in this paper, above all by the rhetoric of assurance that any radical changes were only 'for the duration'. The argument of war emergency won widespread acceptance for the mobilisation of women on such an unprecedented scale, as did those discursive manoeuvres that glamourised and domesticated war industrial work and military service, womanised those sectors of the military and heavy industry in which women were employed, and rehabilitated the image of female service personnel as ladies, not 'loose' women. Despite its wartime reconstitution, including the proliferation of eroticised images of women, femininity was defined so as both to signify and to maintain women's differ ... from, deference towards, and dependence on men. Nevertheless, the fears unleashed by war's destabilisation of gender relations survived beyond the end of the war.

In September 1945, despite the fact that approximately 80,000 women war-workers had already been laid off by then, a popular Canadian magazine ran a cartoon that derived its humour from the preposterous yet feared possibility of a reversal of the sexual division of labour between male bread-winner and dependent female domestic. A husband wearing an apron and standing with a mop in his hand and a bucket at his feet frowns in annoyance at his overalls-and-bandana-clad wife who, home from the factory, has headed straight for the refrigerator, tracking muddy footprints across his nice, clean floor.[56]

NOTES

1 This paper draws heavily on Ruth Roach Pierson, *'They're Still Women After All': The Second World War and Canadian Womanhood* (Toronto: McClelland & Stewart 1986). See also Ruth Roach Pierson, 'Beautiful Soul or Just Warrior: Gender and War', *Gender & History*, i, 1 (Spring 1989), pp. 76–86.

2 Eleanor F. Straub, 'United States Government Policy Toward Civilian Women During World War II', *Prologue: The Journal of the National Archives*, v, 4 (Winter 1973), p. 240.

3 *Saturday Night*, 26 Sept. 1942, p. 10.

4 Lotta Dempsey, 'Women Working on Ships and Aircraft', *Mayfair*, June 1943, p. 74.

5 Notes for the Assistance of Speakers at School for NSS Employment Office Personnel, February 1943 (National Archives of Canada [NAC], Records of the Department of National Defence, RG 24, Reel No. C–5303, File HQS 8984–2).

6 *Saturday Night*, 4 Mar. 1944, p. 4.

7 *Mayfair*, May 1943, pp. 141–5.

8 'The Income Tax Change Applying to Married Employees in 1947', n.d. (NAC, Records of the Department of Labour, RG 27, vol. 606, file 6–21–11).

9 Letter of 30 Ap. 1942 from E.M. Little, Director of NSS, to G.S. Tattle, Deputy Minister, Department of Public Welfare, Ontario (NAC, RG 27, vol. 611, file 6–52–6–1, vol. 1).

10 *Women at War*, comp. and ed. by J. Herbert Hodgins, David B. Crombie, Eric Crawford, and R.B. Ruestis, with cooperation of the Department of Munitions and Supply and the National Film Board of Canada (Toronto 1943), pp. 70–1.

11 Minutes of the 89th Meeting of the National Campaign Committee, 22 Feb. 1943 (NAC, RG 24, Reel No. C–5303, file HQS 8984–2).

12 *Report: An Enquiry into the Attitude of the Canadian Civilian Public Towards the Women's Armed Forces* (Montreal/Toronto: Elliott-Haynes Ltd 1943), hereafter cited as Elliott-Haynes, *An Enquiry*.

13 *Report of Enquiry–Canadian Women's Army Corps: Why Women Join and How They Like It*, April-May 1943 (copy at Department of National Defence [DND], Directorate of History [DH], 168,009 [D91], hereafter cited as *CWAC Report*).

14 Notes for the Assistance of Speakers for NSS Employment Personnel, Feb. 1943 (NAC, RG 24, Reel No. C–5303, file HQS 8984–2).

Canadian Women's Army Corps recruiting stand (Manitoba Archives, Canadian Army Photograph collection, no. 162)

Female armaments worker (City of Toronto Archives, G&M 91963)

Army driver (Manitoba Archives, Canadian Army Photograph collection, no. 157)

Women workers' Beauty Contest (City of Toronto Archives, G&M 80371)

15 *Women in Khaki* (1942), p. 20 (copy at DH, DND, 164.069 [D1]).

16 Lionel Shapiro, *The Sixth of June* (Garden City, NY 1955); Geoffrey Cotterell, *Westward the Sun* (London 1952).

17 Lotta Dempsey, 'They're Still Feminine', *Maclean's Magazine*, 1 Jan. 1943, p. 7.

18 *Maclean's Magazine* 1 Mar. 1942, p. 3.

19 York University Archives, *Toronto Telegram* Print Collection, RCN photo by L.F. Shera-ton, RCNVR; also NAC, National Photography Collection, DND/PAC/PA–141002, unidentified WRCNS signaller wearing bellbottom trousers, Vancouver, Feb. 1944.

20 'C.W.A.C. Winter Driving Uniform', 9 Jan. 1940 (NAC, National Photography Collec-tion, Z–61–2).

21 Emphasis in original. Master-General of the Ordnance Letter no. 64/1944, 22 July 1944 (NAC, RG 24, vol. 2255, file HQ 54–27–111–33, vol. 3).

22 NSS Representatives Conference re Recruitment of Women for Armed Services, London, Ontario, 15–17 Feb. 1943 (NAC, RG 24, Reel No. C–5303, file HQS 8984–2).

23 Mary Ziegler, *We Serve That Men May Fly: The Story of the Women's Division Royal Canadian Air Force* (Hamilton, Ontario 1973).

24 CWAC Recruiting Pamphlet used in MD, no. 4 in Jan.-Mar. 1943 campaign (DH, DND, 164.069 [D1]).

25 Ziegler, *We Serve That Men May Fly*, pp. 66–7.

26 L.S.B. Shapiro, 'They're Still Women After All', *Saturday Night*, 26 Sept. 1942, p. 10.

27 'Suggested Notes for the Guidance of Speakers at the National Selective Service Schools', 9 Feb. 1943, sent out from the Office of Director of Army Recruiting (NAC, RG 24, Reel No. C–5303, file HQS 8984–2).

28 See Ruth Milkman, 'Redefining 'Women's Work': the Sexual Division of Labor in the Auto Industry During World War II', *Feminist Studies*, viii, no. 2 (Summer 1982), pp. 336–72, for a discussion of the flexibility of what Milkman calls 'the idiom of sex typing' as a component of 'the resilience of the structure of job segregation by sex'.

29 See Ruth Roach Pierson, chapter 3: 'CWAC of All Trades', in *They're Still Women After All*.

30 York University Archives, *Toronto Telegram* Photograph Collection.

31 Milkman, 'Redefining Women's Work'.

32 Wartime Information Board Reports Branch, 'Immorality of Service Women', confiden-tial memorandum, 19 Mar. 1943 (DH, DND, PARC, box 4824, S–2730–1/15, vol. 1).

33 A decision was taken in July 1943 'that no published statistics or retorts should be made' in the military's attempts to combat the rumour mongering. J.N. Buchanan, 'Canadian Women's Army Corps, 1941–1946', Report No. 15, Historical Section (GS), Army Head-quarters, 1 May 1947 (Copy at DH, DND).

34 *50 Questions and Answers about CWAC* (Ottawa: King's Printer 1944), p. 11; 'Military Police-women in Big Sister Role', *Globe and Mail*, 5 Dec. 1942, p. 12.

35 Inside cover, *Canadian Geographical Journal*, Dec. 1943.

36 Manitoba Archives, Canadian Army Photography Collection, no. 194.

37 Ruth Roach Pierson, Introduction to chapter 2, 'Sexuality', in Beth Light and R.R. Pier-son (eds), *No Easy Road: Women in Canada 1920s to 1960s* (Toronto 1990), p. 86.

38 Susan Gubar, 'This Is My Rifle, This Is My Gun', in Margaret Randolph Higonnet, Jane Jenson, Sonya Michel, and Margaret Collins Weitz (eds), *Behind the Lines: Gender and the Two World Wars* (New Haven & London 1987), pp. 257–8.

39 Pierson, Introduction to chapter 2, in Light and Pierson, *No Easy Road*, p. 87.

40   One such photograph first appeared in the *Montreal Standard* of Aug. 1943, and was later used as the basis for a CWAC recruiting poster.

41   NAC, National Photography Collection, PA–141005.

42   Some World War II servicewomen do not have pleasant memories of their having been used as sex objects. One wrote that 'as an RCAF (WD) I was involved in displaying my legs on a float in an advertising effort to recruit young men (so-called)· still in high school. What a horrid memory.' Letter to author from Shirley Goundrey, dated 17 Oct. 1977.

43   'Regulations for the C.W.A.C., 1941', Section IX, para. 120(b) (NAC, RG 24, vol. 2252, file HQ 54–27–111–2, vol. 1). 'Regulations for the C.W.A.C., 1942', Section IX, para. 40(b) (DH, DND, 113.3C1 [D1])'.

44   W. Hugh Conrod, *Athene, Goddess of War: The Canadian Women's Army Corps* (Dartmouth, Nova Scotia 1983), p. 55.

45   NAC, National Photography Collection, Caption to Army photo, Negative Z–2337–5, 3 Dec. 1943.

46   NAC, National Photography Collection, Caption to Army photo, Negative Z–2337–2, 3 Dec. 1943.

47   NAC, National Photography Collection, 'CWAC Beauty Culture', Negative Z–2431–3, 8 Jan. 1944.

48   NAC, National Photography Collection, Z–2405–3, 27 Dec. 1943.

49   Rosamond 'Fiddy' Greer's account of life in the Women's Royal Canadian Naval Service during the Second World War contains this personal recollection: 'When two sisters slept together in a lower bunk across from mine, the thought of lesbianism never crossed my mind. This is not at all surprising, as I had never even heard of it. The more worldly-wise than I (which was probably almost everyone) may have known about "it", but none spoke the unmentionable word and for some time I thought how nice it was that the girls could comfort one another when they felt homesick. However, one day I heard that they had been "found out": shortly afterwards they disappeared from *Stadacona*; and I surmised that "something funny" had been going on. I was learning; although not too swiftly … for a long time after the incident I thought lesbians were rather peculiar sisters.' Rosamond 'Fiddy' Greer, *The Girls of the King's Navy* (Victoria, BC 1983), pp. 83–4.

50   NAC, National Photography Collection, Army Photo Negatives Z–1885–17, 11 Aug. 1983.

51   NAC, National Photography Collection, Army Photos, 'C.W.A.C. Wedding', Negatives Z–2586–(1–9), 1 Mar. 1944; 'CWAC–Navy Wedding at Kildare Barracks', Negatives Z–3796–(1–4), Apr. 1945.

52   'Private of C.W.A.C. Weds Army Corporal', *Globe and Mail*, 2 Dec. 1942, p. 11.

53   Wartime Information Board, 'Immorality of Service Women'.

54   As Susan M. Hartmann has written in *The Home Front and Beyond: American Women in the 1940s* (Boston 1982), 'Women's military service represented one of the greatest assaults on traditional practices and values' (p. 212).

55   Thelma Lecocq, 'Woman Power', *Maclean's Magazine*, 15 June 1942, pp. 10–11, 40.

56   Drawn for *Maclean's* by Vic Herman, *Maclean's Magazine*, 15 Sept. 1945, p. 32.

HOPES FRUSTRATED: THE IMPACT OF THE KOREAN WAR UPON BRITAIN'S RELATIONS WITH COMMUNIST CHINA, 1950–1953

Peter Lowe

Before the outbreak of the Korean war in June 1950 British policy towards China aimed at fostering a new positive relationship which would transform the international scene in East Asia and reduce the dangers posed by the threat of monolithic world communism. In 1949 the rapid disintegration of the *Kuomintang* regime, now confined tenuously to its redoubt in Taiwan, connoted the effective victory for the Chinese communist movement led by Mao Tse-tung. The Labour government of Clement Attlee decided to recognise the new administration in Peking and recognition was accorded early in January 1950. The decision was reached for a combination of reasons.[1] The common sense British view was that a government which was in control of most of a country should be recognised providing that it appeared capable of retaining control and of fulfilling the basic criteria applicable in diplomatic relations. Within the British Commonwealth, India was in the forefront in demanding a more positive reaction to the growth of nationalism in Asia and the Attlee government wished to work as closely as was feasible with New Delhi so as to strengthen the evolution of the modern Commonwealth. In addition, British firms in China, particularly in Shanghai, favoured recognition on the basis that this was the only way of ensuring a future for trade. Within the Foreign Office an important consideration concerned the nature of the relationship between the two giants of the communist world. Were Moscow and Peking joined so intimately by ideology as to constitute a malevolent unified force, as maintained by growing numbers of Americans, or was there a possibility that friction between them could lead to an independent communist policy being followed in Peking? The decision to extend recognition inevitably entailed the most significant divergence in Anglo-American relations in East Asia since the end of the Pacific war. The Truman administration was thoroughly weary of Chiang Kai-shek's regime in 1949 as was demonstrated clearly in the publication of the China White Paper in August 1949.[2] However, the foreign policy of the administration was based upon vehement hostility to communism, compounded by the repercussions of the consolidation of communist control in eastern Europe. Truman and his Secretary of State, Dean Acheson, were castigated by right-wing Republicans and by some Democrats for being allegedly 'soft' on communism.[3] Such a charge would have been laughable and could have given rise to hysterical reactions had it not been for the hysteria of a different nature created by the harsh accusations made by Senators Taft, Knowland, Wherry, McCarthy and others. Truman and Acheson had little room for manœuvre and, in any case, the President was strongly opposed to communist governments and had no intention of hastening to recognise one.

Therefore, Britain and the USA went their separate ways in January 1950. Truman and Acheson viewed the decision reached by Attlee and Bevin more in sorrow than in anger. Considerable numbers of Americans reacted with ire and regarded British recognition as a treacherous stab in the back from a socialist government of suspect loyalty.[4] The general assumption within official circles in London and in commercial circles had been that the Chinese authorities would soon adopt a more mellow approach to political and economic relations with Britain. This was a logical and rational deduction to make. It omitted to incorporate an accurate forecast of Chinese reactions based on the motivating influences within Chinese communism. Mao, Chou En-lai, Liu Shao-ch'i, Peng Teh-huai and their colleagues were driven on by fierce detestation of Western imperialism and the manner this had manifested since the 1830s.[5] They believed history was on their side and that world imperialism and capitalism were doomed to extinction. Why should they rush to embrace the British government which called itself 'socialist' yet which worked so closely with Washington? There were differences within the Chinese leadership over the most appropriate attitude to follow, although the full character of this cannot be appreciated without access to Chinese sources. The cosmopolitan Chou En-lai, shrewd and urbane, was willing to contemplate certain dealings with the West on his terms but he did not control policy.[6] Mao was inward-looking, suspicious of the West, and imbued with revolutionary zeal of a somewhat naive nature following the historic achievements of the Yenan era. While Mao had no profound admiration for the Soviet Union, he respected Stalin's successes in transforming the Soviet State and society: given the horrendous economic problems confronting him, Mao concluded that there was no alternative to reliance upon the Soviet Union in the short to medium term.

British diplomats in Peking found themselves ignored by the top Chinese leaders. They could deal with middle or lower grade officials but had no access to the top. For an insight into thinking at this level they had to rely on information retailed by K.M. Panikkar, the intelligent but idiosyncratic Indian ambassador in Peking.[7] Bevin was very disappointed at the negative Chinese attitude. When he met representatives of the China Association in London in the middle of March 1950 he expressed doubts as to whether the cabinet had been correct in the timing of recognition: it might have been better to have waited longer, although there had been pressing reasons for approving recognition when it occurred because Mao had been in Moscow then for negotiations with Stalin and it was wise to demonstrate that the West was not too antagonistic to Mao's government.[8] As regards Chinese membership of the UN, British policy was in part inhibited by American views: when a majority within a UN agency supported admission of the Peking government, this should be accepted. Logically this applied to the far bigger question of the communist government taking over the seat in the UN Security Council occupied by Chiang's regime. Since the United States was wholly opposed to allowing Mao's government into the UN, a difficult dilemma remained. Thus on the eve of the Korean war Anglo-Chinese relations were very cool and the best that could be anticipated was a gradual thaw. Even this was rendered doubtful because of a tougher American stance over Taiwan, which was fundamentally communicated by Acheson to the British ambassador early in June.[9] Britain maintained relations with the regional authorities in Taipeh: the Chinese communists were not persuaded by the cogency of the subtle distinction drawn by the British authorities and regarded it as a further example of perfidy. At the start of the Korean war President Truman issued an important statement in which he explicitly linked the crisis in Korea with the situation obtained in Taiwan: in essence the *status quo* would be frozen and neither the communists nor the *Kuomintang* could attack each other, which meant that the promised communist invasion of Taiwan would not materialise.[10] This action was deplored by the Attlee Cabinet which preferred to treat Korea and China as discrete issues. However, the British consul-general in Tamsui, E.T. Biggs, informed the Foreign Office in August 1950 that he believed that the United States had taken the correct decision on political and moral grounds:

If the United States [is] involved in hostilities in the island's defence and calls for assistance it is in my opinion desirable even at the expense of our relations with Peking that we support her. Furthermore I consider we should make our position clear now. Any impression that Britain would stand aloof and that the Formosa issue can be utilised to divide Britain and America as suggested in some sections of the British press, will only encourage a Communist attack on Formosa and must at all costs be avoided.[11]

Biggs thought that the best solution would be to transfer responsibility for the defence of Taiwan to the UN. Biggs perceived the *Kuomintang*'s presence as a major obstacle but he held that this could be surmounted through the application of sufficient pressure once *Kuomintang* funds were exhausted. Biggs's reasoning was not dissimilar to that motivating Truman, Acheson, and John Foster Dulles (who was handling the negotiations leading to a Japanese peace treaty). The Truman administration did not want to commit itself to propping up the moribund *Kuomintang* but instead to search for a solution under UN auspices, perhaps meeting the aspirations of the indigenous inhabitants of Taiwan who had been brutally suppressed by the *Kuomintang*. The Foreign Office and cabinet did not wish to take any steps that would markedly exacerbate relations with Peking. Anxiety had already been stimulated by General MacArthur's visit to Taiwan on 31 July and 1 August which indicated the dangers of increased intervention that would render the communists even more indignant.[12] The crucial feature was that Taiwan would be denied to the communists whatever happened.

As regards the conflict in Korea it was hoped in London that China would not intervene militarily. The communist government was faced with such grave economic difficulties that it appeared improbable and foolish for China to act. The Attlee cabinet could discern no real reason why China might intervene, for North Korea was held to have attacked at the instigation of the Soviet Union or with its connivance. The UN Command was concerned with defeating the aggressor in Korea and not with invading Chinese territory. Politicians and the civil servants in the Foreign Office were generally complacent. MacArthur's dramatic success in the Inchon landing in September 1950 gave rise to supreme optimism, with a determination to liberate the whole of Korea and thus to roll back communism. A cautionary note was sounded by the British chiefs of staff who feared that China could intervene: as British defence resources were severely stretched in Europe, Malaya, and East Asia, there was no wish to see this extended further.[13] Britain co-sponsored the resolution of the UN General Assembly on 7 October approving an advance north of the 38th parallel with the aim of unifying Korea on a democratic basis. China's reaction had sharpened since late September and Mao was alarmed at the threat to Manchuria implicit in a march by UN forces to the Yalu river. Around the middle of October, Chinese forces crossed into Korea and the first military engagement occurred towards the end of October. Alarm in Britain increased markedly, accentuated by the warnings of the chiefs of staff. The Cabinet proposed a buffer zone in the north of Korea so as to obviate full-scale Chinese action. The Truman administration was unenthusiastic for the buffer zone, since it was still attracted by rollback. Truman did not want war with China but neither was he prepared to curb MacArthur. In the last week of November all doubt regarding Chinese intentions was dispelled by a massive Chinese attack. This effectively ended any prospect for improvement in Anglo-Chinese relations for a long time to come.

In December 1950 Attlee flew to Washington to meet Truman. It was provoked by the President's reference in a press conference to the possible use of the atomic weapon in Korea but Attlee wanted a full exchange with Truman on world problems. Attlee urged the merits of attempting to achieve a settlement that would address the problems of Chinese representation in the UN and Taiwan. Truman was predictably unenthusiastic: as the latest oppressor in Korea China should not be rewarded.[14] The gravity of the position and extent of pressure from perplexed members of the UN led to hectic endeavours within the UN to see if a solution to Chinese issues could be reached. Truman and Acheson reluctantly acquiesced in the cynical but well-founded belief that China would not compromise, believing that it might be able to push its military successes in Korea to greater heights. UN attempts at conciliation petered out in January 1951 and Acheson succeeded in pushing through the General Assembly a motion formally condemning China for aggression in Korea. Deep concern was caused within the cabinet by American determination to increase the scope of measures against China, because it was feared that these could give rise to Chinese action against Hong Kong and could further escalate into all-out war. Britain supported the USA in the vote in the General Assembly on 1 February 1951 but considerable misgivings remained.[15] The British chargé d'affaires in Peking, John Hutchison, who had vainly tried to improve relations with Peking for the previous twelve months, reported on 2 February that the fact that Britain had voted in the UN in favour of condemning China would doom prospects for establishing normal relations. He feared that Chinese officials would become even more difficult to deal with.[16] A.A.E. Franklin and Peter Dalton of the Far Eastern department criticised Hutchison on the grounds that the Chinese would do whatever suited them and there was little point in being naive over this.[17] Urgent consideration to the possible application of sanctions was given within Whitehall in February. A telegram was sent to Washington on 17 February confirming the sombre warning given by Sir Gladwyn Jebb in the General Assembly on 1 February that, 'My Government has the gravest doubts whether any punitive measures can be discovered which are not dangerous, double-edged or merely useless.'[18] The political consequences would be serious in terms of undermining Western contacts with China and would create tension with India and other Asian members of the Commonwealth. In the economic sphere drastic sanctions were opposed because they would have no significant effect on China as a whole; they would accentuate tension in East Asia; would increase the danger to Hong Kong and Indo-China; would not be supported by various members of the UN; and would quite likely split the Commonwealth between the old white dominions and the three Asian members. The telegram continued:

The furthest that the United Nations might reasonably attempt to go in the event of a failure of all chances of a negotiated settlement would perhaps be a selective embargo of strategic materials, so as to limit China's military potential. Even this limited measure will increase tension in the Far East and is unlikely to command the support of the Asian members of the United Nations and in particular of India; though, in the worst event, it may be considered the minimum measure necessary to maintain the dignity and position of the United Nations.[19]

In addition, it was necessary to recall that sanctions against China would be 'playing Russia's game from the political standpoint and will also tend to drive China more and more into the hands of Russia and her satellites for economic and trading purposes'. The government held that the emphasis of work in the UN should be put on the functioning of the Good Offices committee and not on the Study Group on Additional Measures. It was important to leave scope for negotiations with China should the Peking government wish to talk in future. Of course, it was necessary to reveal understanding for the American viewpoint but British reservations must be fully conveyed.

Jebb requested clarification in February 1951 on the current aims of British policy. Robert Scott wrote to him on 17 February setting out the premises of the approach favoured by the Foreign Office. The principal aim was to exert 'a moderating influence on Peking', in the hope that 'normal relations' could be achieved in future.[20] It was difficult to foresee progress until China was convinced that the West could provide something the country needed. Britain wanted to remain in China, difficult as it was:

Our foothold, political, commercial and cultural in China is tenuous, but we are reluctant to see these links completely severed since once we abandon this foothold all Western influence in China will disappear. For the present, therefore, our aim continues to be to pursue a policy of patience, caution and restraint, but at the same time to deal firmly and bluntly with the Chinese authorities if and when occasions arise which call for plain speaking. I entirely agree that any attempts to ingratiate ourselves would almost certainly have exactly the reverse effect of the one desired. The Chinese, as you point out, are tough, shrewd and realist, and more likely to understand straight speaking than is generally assumed to be the case. Our aim should therefore be to adopt a firm but not hostile attitude.[21]

The British government and people regarded the Chinese people in a positive way and appreciated the strength of Chinese civilisation. With some strength of imagination on his part Scott argued that the Chinese and British governments shared the same aim – 'a prosperous, independent, and unified China'.[22] While due acknowledgement should be given to the positive features in China the negative ones must be underlined. It was likely that many Chinese were disturbed by the nature of the relationship with the Soviet Union: it might be apposite when the occasion arose, to remind China of the peril of being used 'purely as a catspaw for Soviet Communism in the East'. Corruption had apparently disappeared and the country was more efficiently run but a police state was clearly developing: the free press had vanished and ominous signs of a regimented approach resembling the Soviet Union's could be seen. Scott's analysis was in part sensible but simultaneously betrayed a characteristically paternalistic air which grated with the Chinese.

The United States desired a complete embargo but was reluctantly prepared to settle for one including petrol, oil, atomic energy materials, arms and ammunition, and weapons of war.[23] The British reaction was to dissuade the Americans from going too far. In Peking Hutchison had his farewell meetings with Chinese officials before departing for retirement. It was a sad occasion given the optimistic tone of a year before. Hutchison stated that he had been greeted with courtesy but without warmth: an attempt on his part to discuss progress in diplomatic relations had been rebuffed.[24] His successor was Lionel (Leo) Lamb. The Foreign Office advised Lamb on how to conduct himself in his initial meeting with Chinese officials in a telegram sent on 2 March. This was couched in terms of the opinions conveyed in Scott's letter to Jebb referred to earlier. Lamb was advised to state regret at the tendency of Chinese leaders to cite frequently 'the insincerity of His Majesty's Government's policy towards China'.[25] A week later Lamb reported that he had seen a vice-minister for foreign affairs the previous day: they had exchanged the usual courtesies but the vice-minister evaded Lamb's endeavour to pursue concrete political matters. The vice-minister referred favourably to the businessman, John Keswick, but it was hard to say what significance was connoted by the remark. The stalemate in relations continued.

Much American criticism of Britain centred upon goods reaching China via Hong Kong. This in turn demonstrated a broader American feeling that the colonial authorities in Hong Kong were unduly conscious of the undesirability of offending China as further illustrated in a controversy over the fate of several aeroplanes impounded in Hong Kong pending resolution of a legal dispute as to whether they belonged to the communist government or to a company associated with the *Kuomintang* and financed by the CIA.[26] In January the governor of Hong Kong, Sir Alexander Grantham, warned that the effects of sanctions on the crown colony would be very serious. If China retaliated, industries in Hong Kong would suffer; trade would be reduced by up to 50 per cent of exports via usual channels and by a possible 20 per cent of exports via unofficial channels (smuggling). Unemployment would ensue, itself giving rise to growing unemployment. Shipping would suffer and a spiral of despondency could develop.[27] A.A.E. Franklin minuted that a powerful body of American opinion was critical over Hong Kong on the grounds that those in the colony were simply interested in making money and did not worry about political responsibilities. He noted that export of all strategic materials to China was banned in which respect China was treated in the same way as the Soviet Union and eastern Europe. Oil, machine tools, defined chemicals, munitions, military supplies, planes and ships were prohibited. In the first nine months of 1950 Hong Kong exports to China and Macao were worth a figure of £65.9 million; imports amounted to £41.2 million. Franklin observed that American officials were most concerned about rubber, drugs, metals, trucks, steel plates, and raw materials including cotton.[28] Clearly it was important to reassure Washington, and the Foreign Office felt that it was reasonable for the American consul-general in Hong Kong to be given information regarding re-export of American goods to China. One difficulty was the State Department's demand that no goods manufactured from United States supplies should be exported to China. This would have to be discussed with the Americans.[29] Grantham responded that the consul-general had access to considerable information and that he was opposed to providing additional facilities.[30] The Foreign Office was critical of the Hong Kong authorities, deeming them to be evasive. A rather querulous telegram from Grantham, sent on 13 February, provoked N.C. Trench of the Foreign Office into commenting that he would have greater sympathy with Grantham's indignation if there were not grounds for thinking that officials in the colony had made no serious effort to render export controls effective; in addition, they had dealt ineptly with the Americans.[31]

The Foreign Office wished to go a reasonable distance towards meeting legitimate American concern but it was necessary to remind the Truman administration of the genuine problems encountered in Hong Kong. Policy over sanctions was coordinated via the Far Eastern (Official) Committee, which represented all interested government departments. On 1 March a telegram was sent to Washington stating that Britain was anxious to achieve a satisfactory understanding but alluding to the dangers of possible Chinese action against Hong Kong, which could include an actual occupation of the colony.[32] It was unlikely that such an extreme move would occur but it was politically wise to remind the Americans of the contingency. However, the opinion formed in Washington was that Britain was reluctant to go very far in the application of sanctions and the embassy in Washington reported that the position was most embarrassing.[33]

In April the dismissal of MacArthur caused intensified debate over British policy towards the Korean war and China.[34] MacArthur and his supporters condemned British policy for restraining justified American initiatives and various allegations were made regarding Hong Kong. MacArthur stated that significant supplies were reaching China through the colony. Sir Oliver Franks, the British Ambassador, warned that this would stimulate added criticism of Britain for assisting the enemy in Korea. In order to counter-attack Franks requested up-to-date statistics concerning Japanese exports to China. This was important because MacArthur remained as SCAP in Japan until his removal: therefore, the general's sanctimonious remarks might appear a shade hypocritical once the statistics were revealed.[35] The figures were provided promptly and interestingly showed that the monthly values of Japanese exports to China during the first three months of 1951 were:

January – 493,000 US dollars
February – 827,000 US dollars
March – 1,063,000 US dollars

The chief items comprised textiles, sewing machines, and bicycles.[36] The report added that most of the 16 million dollars of Japanese exports to China in 1950 comprised iron and steel manufactured goods.[37] Franks discussed it with the State Department. The United States was allowing Japan to continue trading in non-strategic goods including cotton fabrics, bicycles, woven silk, sewing machines, cotton yarn, filament, and photo-printing paper. Raw materials and semi-processed goods were banned in December 1950. The State Department confirmed that out of a total of 19 million dollars worth of Japanese exports to China in 1950 approximately 15 million dollars worth consisted of iron and steel goods.[38]

A lengthy debate over the application of sanctions took place in the House of Commons on 10 May. Churchill, as leader of the opposition, emphasised the priority of working closely with the USA and of not allowing the Anglo-American relationship to be disturbed. He stated that over recognition of Mao's government he had supported *de facto* recognition in the latter part of 1949 because this conformed with reality but he had not advocated *de jure* recognition. The Attlee government had accorded full recognition and found itself in the anomalous situation of dealing with a government with which it was at war in Korea. Churchill said that he had not wanted UN forces to advance in the far north of Korea in October-November 1950 and that a broad 'no-man's land' should have been retained between the UN front and the Yalu river.[39]

In the course of May 1951 the Far Eastern (Official) Committee moved to tighten economic measures. Rubber should be added to the list and export licences should be introduced in Britain for all other exports to China. Provision should be made to stop evasion in trans-shipping.[40] The governor of Hong Kong protested that firmer control of trans-shipping would damage the colony's economy.[41] New regulations were approved and introduced in June 1951. The Foreign Office told the State Department that this was a 'generous interpretation' of the UN resolution carried on 18 May. This was as much as Britain could do in existing circumstances.[42] The State Department expressed appreciation. The difficulties inherent in controlling trans-shipment and carriage of prohibited goods were emphasised. The State Department argued that the terms of the UN resolution covered this aspect, so that renewed pressure could be expected. Furthermore American officials wished to see a detailed list of all prohibited goods; this would help other countries and customs officers.[43] The Foreign Office dissented from the interpretation that the UN resolution included trans-shipment. Arguments continued from time to time between Britain and the United States as was unavoidable. In November 1951 statistics produced following allegations made by General Omar Bradley regarding seaborne trade with China revealed that while trade with China had increased markedly after the start of the Korean war a sharp decline occurred during 1951.[44] Given the potent reactions to Chinese action in Korea in the USA, divergent attitudes were inevitable. While differences existed, they were kept in check; however, right-wing Republicans harboured a continued sense of grievance and this surfaced with some frequency in their statements down to and including the presidential campaign in 1952. General Eisenhower, the Republican candidate, did not share the resentment but the political demands of placating each section of the party meant that it could affect some of those who would influence policy-making after January 1953.

Part of the argument in favour of recognition in 1949 concerned the future of British firms in China. British enterprise enjoyed a prominent position in the China trade for over a century even though that dominance had been progressively eroded by German, French, American, and Japanese competition since the end of the nineteenth century. Although there was a certain apprehension, the prevailing feeling was that there would be a future for British commerce in China after an initial period of turbulence. However, the Chinese economy was so backward, accentuated by the ravages of war and civil war, that foreign trade was not a priority for the new government. Passionate feelings regarding the evils done by Western imperialism in the past propelled the Peking government into a tougher stance. China would do little to encourage Western companies to remain and would be glad to see at least some of them depart. If they stayed they would literally pay more for the privilege. The Chinese Labour movement was encouraged to press substantial wage demands and to take industrial action of conventional and unconventional means to secure their aims: this could include imprisoning managers in their offices until demands were met. Chinese officialdom and taxation exerted further pressure. Chinese ire was increased by the fighting in Korea. The prospects for British firms were bleak. A memorandum was compiled in the Foreign Office in April 1952 examining the trends in trade since 1950. This showed that British exports to China were valued at £3.6 million in 1950 and this fell to £2.7 million in 1951. The decline in the latter half of 1951 was sharp. It reflected the tightening of strategic export controls as enforced since June 1951. The decline continued in the first two months of 1952 when the respective figures were £75,000 and £42,000.[45] The chief items comprised copper wire, lead sheet, machinery (including appreciable textile machinery), iron and steel goods, wool tops, chemicals, drugs and dyes, and vehicles including tyres, tubes, and pedal cycles. Apart from wool tops, the main textile export of note was cotton sewing thread. Considerable re-exports of British goods to the Chinese mainland occurred but these could be identified in the Hong Kong returns. Strategic controls would have affected matters in this context. British imports from China were valued at £10.3 million in 1950 but fell to £7.7 million in 1951. The decline resulted from various causes including surpluses being diverted to the Soviet bloc (from which Britain made some purchases), diversion to home consumption, the effects of barter restrictions being imposed on trade, and the problems experienced by British firms. British imports included eggs (not shell), maize, bristles, hair, feeding stuffs, oil seeds, and other oil-based commodites. The British government welcomed increased business in textiles, none of which were regarded as strategic. The USA had severed all direct trade with China and might press Britain further. Chemicals could well be objectionable for strategic reasons and care must be exercised. Additional coal purchases would be of little use to Britain unless a decline could be effected in British shipping calling at Chinese ports. Other commodities likely to be offered by China – basic foods or industrial raw materials – were unlikely to violate strategic controls.[46]

If China proposed a new understanding on trade it would offer a useful market for textiles. A difficulty could be that Britain might be required to accept the supply of strategic materials to China from eastern Germany or elsewhere. This could offend the USA. A possible explanation for the recent Chinese offer of encouraging trade, apart from the manifest purpose of propaganda, would be that China had discovered that its policy of fostering trade with the Soviet bloc resulted in poor prices for exports and would limit the range of purchases unduly. While the Korean war continued it would be difficult to expand exports of capital goods to China but when the war terminated, this could be significant.[47]

The flurry of interest in April 1952 resulted from an initiative taken by certain left-wing British businessmen, including Sydney Silverman, the Labour MP for Nelson and Colne, to encourage trade through negotiations at an international economic conference in Moscow.[48] The talks between the British businessmen and Chinese officials centred around textiles, chemicals and metals in return for which China would sell to British coal, bristles, frozen eggs, and other defined produce.[49] Information indicated that chemicals and metals would be supplied by East Germany with the British contribution comprising textiles.[50] Despite the doubts surrounding the nature of the Moscow deliberations, Anthony Eden and other ministers welcomed the apparent opportunity to encourage trade, which should be conducted through established British merchants in Shanghai and Hong Kong.[51] Anxiety within the government was promoted by the decision announced in the beginning of April by Butterfield and Swire and by Jardine, Matheson and Company that their ships would not visit Chinese ports in future. This was another example of the diminishing role of the traditional firms. The China Association was profoundly concerned and it was decided that Leo Lamb, the minister in Peking, should make urgent representations in Peking for a new positive policy to be adopted regarding British firms. If Lamb's representations failed, then British firms would all act to notify their Chinese representatives of closure.[52] The permanent under-secretary, Sir William Strang, minuted for Eden that the implications were so grave that the Economic Policy Committee of the cabinet should review matters; Eden agreed.[53] Robert Scott observed on 9 April that the members of the China Association were divided over tactics but were in agreement on the principle of withdrawing from the China trade. He recommended that Eden should secure the agreement of his colleagues over the decision to withdraw and to suggest that the decision on the form of the first approach to the Chinese government should be left to his discretion.[54]

On 7 April Eden met a small delegation of leading representatives of British companies interested in China including J.K. Swire and W.J. Keswick.

Eden sympathised with the dilemmas facing them. 'He described it as a terrible chapter in our history and said that in view of the seriousness of the issue he wanted to inform the cabinet.'[5] Those present explained that they wanted Lamb to submit a note underlining the problems faced by British firms couched in unobjectionable terms. The latter was desirable so as to obviate unhappy repercussions for shipping firms which had decided to close down. Companies would then submit individual applications around the end of April expressing their desire to end operations and apply for exit permits for British staff. A major difficulty in leaving China involved the risk of blackmail in the sense of local managers being detained in Shanghai as hostages while large sums of foreign money were remitted. Could the government act so as to control remittances? The assets of Jardine, Matheson amounted to £20 million which they might be forced into handing over so as to obtain the release of British personnel. Butterfield and Swire would have to contemplate handing over property worth around £10 million. A strong Chinese reaction could be anticipated when the firms determined to send no more remittances, which could be about the end of June.[6] The British-American Tobacco Company had negotiated for two years to end operations and had reached a provisional agreement with the Chinese for taking over their business. The three British banks would need longer because of complexities linked with sums held in New York and because of the issue of repayment of pre-war Chinese deposits.

No progress was made in discussions in Peking and the relevant departments of the British government considered what action could be taken. J.K. Drinkall of the Foreign Office gloomily commented that the weapons in our hands were few and double-edged.[7] Matters proceeded at the customary slow pace. On 8 July Lamb reported that the Chinese government had replied officially to earlier representations. He regarded the tone as moderately encouraging: the gist was less acerbic than usual and was not convincing. He thought that the Chinese might be wondering whether the exchange of goods under the much advertised agreement at the Moscow economic conference could be effected without the support of the British government or without the expertise possessed by British firms established in China. Lamb held that better grounds existed for arguing for more equitable terms and not letting assets be sold off too cheaply.[8] The Chinese continued much as before and the hopes briefly entertained in April of an improvement in relations were not borne out. Jardine, Matheson followed the earlier precedents set up by British-American Tobacco and China Soap and agreed to surrender all the company's local assets against remission of liabilities; the local manager believed that assets would be greater than liabilities to the tune of approximately £160,000.

The Chinese authorities would not consider a more lenient arrangement. The company's liabilities included paying all staff and workers and claims in court by government corporations for compensation for an egg contract, sterling deposit and short delivery of wire rods. Lamb irately described the terms as sheer robbery. I.M. McKenzie of the Foreign Office wrote, 'I think that this surrender by Jardines means that no firm will be able to negotiate better terms than assets against liabilities'.⁵⁹ British firms in Tientsin reached the same rueful conclusion in November: the qualified optimism of six months before had vanished.⁶⁰

In December 1952 Eden sent a strong protest to Peking regarding further seizures of property belonging to four companies – the Shanghai Electrical Construction Company, the Shanghai Waterworks, the Shanghai Gasworks, and MacKenzie and Company. The requisition was carried out by the Shanghai Municipality Military Control Commission. Eden demanded to know whether requisitioning was temporary or permanent. Adequate compensation was expected – 'His Majesty's Government wish to draw the attention of the Central People's Government to the fact that this is the third occasion during the past 18 months on which the Central People's Government have taken action of this sort in respect of British interests and that Mr Lamb has been given no assurances in response to his Notes of the 9th May 1951, and 30th September 1952'.⁶¹ The two previous instances involved the Shell Company's property in China and the property of Mellers Shipbuilding and Engineering Works Ltd.

The Korean war came to an end with the signing of an armistice at Panmunjom at the end of July 1953. It was a bitter conclusion to a bloody and savage conflict; it was improbable that a peace settlement of definitive character could be attained for many years to come. The Korean war greatly accentuated the already existing gulf between the West and China. The American action in preventing a communist invasion of Taiwan, the huge Chinese military action in Korea, the harsh and repressive policies implemented by China, and the process of inexorably squeezing out British business combined to doom any possibility of a significant improvement in Anglo-Chinese relations. The original British aspirations of recognition proving to be a bridgehead from which a positive relationship between China and the West could be constructed was too sanguine and underestimated the factors keeping China in relative isolation. Relations would, however, have been less frigid but for the repercussions of the Korean struggle. China was forced to rely more heavily on the Soviet Union for support in the short term, thus achieving the opposite of what had been hoped for by Attlee and Bevin in January

Hopes Frustrated

1950. A Sino-Soviet rift was to emerge by the end of the 1950s but it was not until the early 1970s that a rapprochement between the United States and China was achieved. The era of the Korean war can only be seen as a dismal, negative and frustrating experience in Anglo-Chinese relations.

NOTES

1. For discussion of British policy towards China in 1949–50, see R. Ovendale, 'Britain, the United States and the Recognition of Communist China', *Historical Journal*, xxvi (1983), pp. 139–58, and Peter Lowe, *The Origins of the Korean War* (London 1986), pp. 98–121.
2. For discussion of American policy towards China, see D.M. Borg and W. Heinrichs (eds), *Uncertain Years: Chinese-American Relations 1947–50* (New York 1980), and N.B. Tucker, *Patterns in the Dust: Chinese-American Relations and the Recognition Controversy, 1949–1950* (New York 1983).
3. See R. Koen, *The China Lobby in American Politics* (paperback edn: London 1974).
4. The private papers of Senator Robert A. Taft, Sr, contain numerous examples of irate, ignorant and sometimes semi-coherent letters castigating British policy. See Taft papers, Library of Congress, Washington DC.
5. For assessments of Chinese Communist policies, see J.K. Fairbank and A. Feuerwerker (eds), *The Cambridge History of China*, vol. xiii, part 2 (Cambridge 1986), pp. 609–870.
6. See Dick Wilson, *Chou: The Story of Zhou Enlai, 1898–1976* (London 1984), for a useful portrait.
7. Panikkar was able but was regarded by British officials as mercurial and erratic.
8. See record of meeting with representatives of the China Association, 16 March 1950 (Public Record Office, Foreign Office papers: FO 371/83344/30).
9. See letter from Franks to Dening, 7 June 1950 (FO 371/83320/9).
10. See *Foreign Relations of the United States, 1950*, vol. vii, pp. 186–7, Acheson to the embassy in London, 27 June 1950.
11. Tamsui to FO, 20 August 1950 (FO 371/83298/58G).
12. For MacArthur's account of his visit, see Douglas MacArthur, *Reminiscences* (paperback edn: Greenwich, Conn. 1965), pp. 385–6.
13. For discussion of China's role in Korea in 1950, see the articles by A. Farrar-Hockley, P.N. Farrar and Peter Lowe in James Cotton and Ian Neary (eds), *The Korean War in History* (Manchester 1989), pp. 4–10, 66–79, 80–99.
14. See Peter Lowe, 'An Ally and a Recalcitrant General: Great Britain, Douglas MacArthur and the Korean War, 1950–1', *English Historical Review*, vol. cv, no. 416 (1990), pp. 636–7.
15. See Lowe in Cotton and Neary, *The Korean War*, pp. 92–3.
16. Peking to FO, 2 February 1951 (FO 371/92233/16).
17. Minutes by Franklin and Dalton, 6 February 1951 (ibid).
18. FO to Washington, 17 February 1951 (FO 371/92234/33).
19. Ibid.
20. Letter from Scott to Jebb, 17 February 1951 (FO 371/92234/34).
21. Ibid.
22. Ibid.
23. Washington to FO, 23 February 1951 (FO 371/92234/45).
24. Peking to FO, 26 February 1951 (FO 371/92234/48).

225

25   FO to Peking, 2 March 1951 (FO 371/92235/58).

26   This protracted issue gave rise to strong criticism of the line adopted by the authorities in Hong Kong. A central personality in pressing the *Kuomintang* case was General Claire L. Chennault. Interesting correspondence is to be found in the Chennault papers, Hoover Institution, Stanford, California.

27   Grantham to Colonial Office (CO), 9 January 1951, communicated to FO (FO 371/92273/30/G).

28   Minute by Franklin, 15 January 1951 (FO 371/92274/37).

29   FO to Washington, 30 January 1951 (FO 371/92274/59).

30   Grantham to CO, 2 February 1951 (FO 371/92274/71).

31   Minute by N.C. Trench, 16 February 1951, on telegram from Grantham to CO, 13 February 1951 (FO 371/92275/98).

32   FO to Washington, 1 March 1951 (FO 371/92276/111).

33   Washington to FO, 5 March 1951 (FO 371/92276/123).

34   For a full discussion of British views regarding MacArthur, see Lowe, 'An Ally and a Recalitrant General'.

35   Washington to FO, 7 May 1951 (FO 371/92278/165).

36   FO to Washington (ibid.).

37   Ibid.

38   Washington to FO, 18 May 1951 (FO 371/92279/184).

39   *Parliamentary Debates, Commons*, 10 May 1951, cols. 2165–72, as contained in FO 371/92279/176.

40   Minute by M.J.M. Paton, 22 May 1951 (FO 371/92279/196B).

41   Grantham to CO, 5 June 1951 (FO 371/92280/224).

42   FO to Washington, 14 June 1951 (FO 371/92281/228).

43   Washington to FO, 16 June 1951 (FO 371/92281/245).

44   Facts and figures to counter General Bradley's statement, November 1951 (no precise date) (FO 371/92287/357).

45   Memorandum by C.J. Homewood, 21 April 1952 (FO 371/99318/31).

46   Ibid.

47   Ibid.

48   Moscow to FO, 9 April 1952 (FO 371/99318/5).

49   Moscow to FO, 9 April 1952 (FO 371/99318/10).

50   Moscow to FO, 10 April 1952 (FO 371/99318/11).

51   FO to Peking, 17 April 1952 (FO 371/99318/14).

52   Minute by J.K. Drinkall, 2 April 1952 (FO 371/99285/104).

53   Minutes by Strang, 3 April, and Eden, 4 April 1952 (ibid).

54   Minute by R.H. Scott, 9 April 1952 (FO 371/99285/108).

55   Record of meeting in Eden's room, House of Commons, 7 April 1952 (FO 371/99285/14).

56   Ibid.

57   Minute by Drinkall, 2 May 1952 (FO 371/99287/153).

58   Peking to FO, 8 July 1952 (FO 371/99290/256).

59   Peking to FO, 3 November 1952 (FO 371/99296/400).

60   Minute by J. Snodgrass, 7 November 1952 (FO 371/99296/403)

61   FO to Peking, 10 December 1952 (FO 371/99346/51).

VIETNAM: THE WAR AMERICA TRIED TO LOSE ?

Stephen G. Rabe

Since 1982 I have been teaching an undergraduate course, 'The American Experience in Vietnam', at the University of Texas at Dallas. In this course we survey the political, diplomatic and military characteristics of the war through lectures, monographs, memoirs and videotapes. Students read incisive surveys like George Herring's *America's Longest War* and Stanley Karnow's *Vietnam* and examine first-person accounts of the war like Philip Caputo's *A Rumor of War*, Robert Mason's *Chickenhawk* and William Broyles's *Brothers in Arms*.[1] Students also view documentaries such as *Vietnam: a History* (1984), a US Public Broadcasting System series, *Why Vietnam?* (1965), an official defence of the war, and *Hearts and Minds* (1975), a bitter attack on the war. This is, of course, the traditional approach to the teaching of history.

But, during the academic semester, we also review how Americans have interpreted and reacted to the war by reading novels, plays, and poetry and listening to music contemporary to the Vietnam era. Examples of this material include award-winning novels such as *Paco's Story* by Larry Heinemann and *Fields of Fire* by James Webb, and the plays of Tom Cole and David Rabe.[2] In addition, we survey how Hollywood has interpreted Vietnam by viewing feature films ranging from the *Green Berets* (1968) to *Apocalypse Now* (1979), *Platoon* (1986) and *Hamburger Hill* (1987).

What makes this course both a special and intense experience is not the material, but the students who enrol. The University of Texas at Dallas (UTD) is a multi-purpose public university in a major metropolitan area. A commuter campus, UTD has a diverse student population of approximately 9,000. The median age of the students is twenty-eight years of age. It is not unusual to have in class a forty-year old sitting next to an adolescent. Accordingly, enrolled in the class will be Vietnam veterans, spouses of veterans, relatives of deceased veterans, anti-war activists, Indochinese refugees, and even now the children of veterans. Vibrant discussions and debates naturally occur in such a mix. The class always has a healthy enrolment and has attracted notice in the Dallas media. Courses on Vietnam are similarly popular throughout the United States.

I came to teach what has become my favourite course because of a question that I was asked during the fall of 1980, the presidential campaign season, in my survey course on US foreign relations. That question serves also as the focus of this chapter. The question asked by a young undergraduate was: 'Why did the United States not try to win the war in Vietnam?' That question, which I was repeatedly asked throughout the 1980s, both angered and startled me.

The question angered me because of my personal background. As a young adult in the 1960s and 1970s, I opposed the war in Vietnam. Like other young men of conscription age, I faced the prospect of being drafted and sent into combat. My decision was to join the Marine Corps Reserve, a choice that virtually guaranteed that I would not be sent overseas.[3] But at the same time, I worked hard and peacefully to change US policy in Vietnam. The moral and ethical questions apart, I thought then and I believe now that intervention in Vietnam was not in the best interest of the United States. Vietnam was a peripheral concern of the United States, not a vital one, and the United States would exhaust and weaken itself seeking unattainable and perhaps undesirable goals. As Walter Lippmann, the influential American journalist who helped coin the term 'Cold War' once noted: 'No nation however strong has universal world power which reaches everywhere ... and no foreign policy is well conducted which does not recognize these invincible realities.'[4]

I suppose that initial question also bothered me in a visceral way. During my anti-war years, I never thought anyone would ever think that there was not enough bombing. Indeed, I despaired that the bombing and killing would ever stop.

The student's question in 1980 surprised me, because I had followed Gerald Ford's advice. In mid-1975, following the collapse of the government in Saigon and the reunification of the country under the control of Hanoi, President Ford was asked at a news conference what were the lessons of Vietnam for the United States. Ford replied 'that the lessons of the past in Vietnam have already been learned – learned by Presidents, learned by Congress, learned by the American people – and we should have our focus on the future'.[5] In essence, Ford's answer was a call for a national amnesia – to forget the whole sordid, bitter business. At the time, his answer seemed to strike a responsive chord among the public. One survey of US history textbooks used by secondary-school students found that the war in Vietnam merited on average about six paragraphs of description.[6] I gave only one lecture on Vietnam in my undergraduate course on US foreign relations and was rarely asked any questions. I knew a revisionist thesis was arising about Vietnam, but, until the early 1980s, I did not think anyone would take it seriously.

A fair sampling of 'Vietnam Revisionism' would include the following works. *Strategy for Defeat* by Admiral US Grant Sharp and *A Soldier Reports* by General William Westmoreland are two influential military post-mortems. Sharp commanded the US Pacific fleet during the war and Westmoreland was from 1964 to 1968 the commander-in-chief of US forces in Vietnam. *Big Story*, by journalist Peter Braestrup, is a critical study of the media's coverage and interpretation of the Tet Offensive of 1968. And two fiery critiques of the diplomatic and political conduct of the war are *Why We Were in Vietnam* by Norman Podhoretz and *No More Vietnams* by Richard Nixon.[7] Although these and other authors emphasise particular themes, they have a common agenda. They argue that the United States should have and could have preserved an independent, non-communist South Vietnam. In short, they can be dubbed the 'Win School.'

The interpretations of the Win School can be summarised. These authors passionately argue that the United States did not violate international law or morality in the tactics and strategies pursued during the war. Indeed, protecting people from communism is a noble cause. But the United States failed to use its military power effectively to accomplish its legitimate goal. The policy of gradual escalation was wrong. Once committed, the United States should have launched massive air attacks on North Vietnam, as was belatedly done during the Christmas season of 1972. Moreover, politicians and officers imposed too many restricted fire zones, handicapping ground operations against the *Viet Cong*. The United States should have also blockaded Cambodia and Laos, cutting off communist supply lines, and perhaps invaded North Vietnam. The use of nuclear weapons should have been kept open as an option. The aim of war is to win. But, as popular wisdom now holds, the United States fought the war 'with one hand tied behind its back'.

The Win School has an explanation for why the United States did not bring its awesome power to bear against the enemy. Domestic critics undermined the US war effort and encouraged Hanoi. Because of the noisy opposition to the war, the United States weakened at critical times, such as in the aftermath of the Tet Offensive of 1968. And Congress, responding to the anti-war movement, tied the hands of the executive branch after the Paris Accords of 1973, dooming South Vietnam. At worst, 'a stab in the back' thesis has developed, blaming defeat on student protesters, the press and electronic media, and liberal politicians.

To be sure, others of this school reject such crude implications. Dispassionate analyses of the war, such as *America in Vietnam* by Guenter Lewy, *On Strategy* by Colonel Harry Summers, and *The 25-Year War* by General Bruce Palmer, assign responsibility for failure to US civilian and military policymakers.[8] These authors argue that the United States needed more adept pacification and counter-insurgency programmes. US leaders should also have instructed the public that the defence of South Vietnam, as it developed in South Korea, would entail a lengthy political and military commitment.

The Vietnam Revisionists undoubtedly want to set the historical record straight. But their concerns extend beyond scholarship. As Guenter Lewy candidly wrote, 'Vietnam continues to haunt our minds and continues to exert a powerful influence on our conception of ourselves as a nation and of our role in the world'. Americans must assuage the mythology of guilt and prove they did not try to win 'for our own self-confidence, our moral strength, and our future capacity to act responsibly in world affairs'.[9] Resolve, as Ronald Reagan did during the 1980 presidential campaign, never to enter a war we did not intend to win. Americans must, as President George Bush advised, bury the 'Vietnam Syndrome'. As one critic of this school noted, the Vietnam Revisionists are restudying the old war in order to plan for the new war.[10]

When confronted with questions about the US military effort in Vietnam, my first response to students was to recite statistics. US war planes dropped three times more explosives on North and South Vietnam, Cambodia, and Laos than were dropped by all combatants during World War II. The United States dropped more tonnage on Hanoi and Haiphong during the Christmas campaign of 1972 than Nazi Germany did on Great Britain between 1940 and 1945. The United States spent $120 billion on the war and incurred massive financial obligations for the future in veteran-related expenses and interest on the national debt. The nation also suffered approximately 350,000 casualties, with some 58,000 servicemen and women dying. Estimates of North and South Vietnamese casualties, both civilian and military, are as high as 2 million dead and 4.5 million wounded.[11] An obvious question arises for the revisionists. How much more blood and treasure should have been expended?

In the use of herbicides, the United States dropped over 19 million gallons of Agents Orange, White, and Blue, poisoning millions of acres of land and destroying over 35 per cent of South Vietnam's hardwood forests. In 1985, the International Union for Conservation of Nature and Natural Resources, an international group based in Switzerland, concluded in a study that one-third of the country is 'now considered wasteland'. But the long-term effects may be far more serious. The study noted 'more than 12 years after the spraying, the forests have never recovered, fisheries remain reduced in their variety and productivity even in coastal waters, wildlife has not returned, cropland productivity is still below former levels and there is a great increase in toxin-related diseases and cancer'. Little wonder that the sardonic motto of the teams who sprayed herbicides was 'Only You Can Prevent Forests'.[12]

Beyond these grim statistics, the question remains: did the United States pursue a 'no win' policy? Key political, military and intelligence officers favoured the strategy of slowly building US troop strength to 540,000, gradually escalating the bombing campaign, and fighting a war of attrition through 'search and destroy' missions. US officials presumed that the enemy would break under this precisely calibrated pressure.[13] (Why the North Vietnamese and *Viet Cong* did not succomb is, of course, a vital question.) But would a more rapid deployment of men and material, as advocated, for example, by General Earle Wheeler of the Joint Chiefs of Staff, have altered the outcome?

Major Mark Clodfelter, a military historian at the Air Force Academy, has argued that 'Rolling Thunder', the 1965–8 bombing campaign against North Vietnam, was aimed at a society and enemy that did not exist. North Vietnam was not Germany or Japan; its war-making capacity was not a by-product of its industrial infrastructure, as Clodfelter observes, 'attacks on oil storage areas and electric power plants had a marginal effect on Hanoi's war effort'. For example, northern trains ran on coal or wood rather than oil. The Vietnamese were used to living by candlelight or oil lamps. In any case, until 1972, the majority of the communists fighting in the South were indigenous southerners, the *Viet Cong*. Clodfelter estimates that they obtained up to ninety per cent of their supplies from within the South. Attacks on transportation-related targets, which comprised most of 'Rolling Thunder's' missions, would scarcely reduce the *Viet Cong*'s fighting abilities.[14]

Would more US troops in South Vietnam have produced victory? Because its population had grown rapidly in the mid-twentieth century, North Vietnam, with a population of 18 million, probably could have matched every new 100,000 US troops deployed to the region. An exasperated President Lyndon Johnson caught the dilemma when he asked General Westmoreland: 'When we add divisions can't the enemy add divisions? If so, where does it all end?'[15] In fact, during his comprehensive review of policy in the aftermath of the Tet Offensive, Secretary of Defense Clark Clifford found that the Joint Chiefs of Staff could not specifically tell him how many troops would be needed to defeat the communists.[16]

This debate over the appropriate number of US troops to send to Vietnam erupted because of disputes over the nature of the war. The Johnson and Nixon administrations repeatedly told the American public that the root cause of the conflict was the North's aggression against the South. The *Viet Cong* were an unrepresentative, violent minority. Military authorities, with the approval of President Johnson and General Westmoreland, informed Congress in 1967 that the communist guerrillas in South Vietnam numbered 285,000. The Central Intelligence Agency (CIA) estimated, however, that the guerrillas numbered 480,000.[7] The Johnson administration rejected the CIA's estimate, because it implied that that war was a civil or 'people's war'.

The revisionists have held that the United States should not have ruled out the option of invading North Vietnam. They presume that China, still beset by the turmoil of the Cultural Revolution, would have been incapable of responding, as they had following the US invasion of North Korea in 1950.[8] This is a clear case of trusting in hope over experience. And, as in the case of the Rolling Thunder bombing campaign, the invasion scenario assumes that a solution to the war in the South could be found in the North.

The use of nuclear weapons similarly risked a major escalation of conflict. In mid-1969, President Nixon ordered the National Security Council to develop a plan, code name 'DUCK HOOK', for 'savage, punishing blows' against North Vietnam. The plan permitted possible use of nuclear weapons under 'controlled' situations. The study group, chaired by Henry Kissinger, concluded that such attacks would probably neither intimidate Hanoi nor significantly reduce the North's capability to wage war in the South.[9] Whether Beijing or Moscow would have responded to nuclear attacks on its client state is an ominous uncertainty.

The revisionists have also argued that the United States should have deprived the North Vietnamese of their sanctuaries in Cambodia and Laos. With operations such as the US incursion into Cambodia in April 1970, the war was brought to Vietnam's neighbours. But the effect was to widen the war, not shorten it.[20] Moreover, the permanent patrolling of South Vietnam's 900 mile border would be logistically difficult and expensive. For example, between 1954 and 1989, the United States spent $100 billion on maintaining 50,000 troops in South Korea.[21]

The thrust of critiques offered by the Vietnam Revisionists is that the United States did not fully use its awesome technology, that it did not pursue 'total' war. US forces could have cleared the towns and villages and destroyed them. The entire country could have been declared a 'free-fire zone'. But as Vietnam veteran William Broyles has archly observed: 'This logic works only if the people are the enemy. And if they were, then whom were we fighting for?'[22] In any case, the revisionists are wrong; restraint did not always characterise US military operations. In retaking the imperial city of Hue during the Tet Offensive, US Marines destroyed this cultural centre of Vietnam. Under rules of engagement, field commanders, once they had secured approval from superiors, were permitted to respond to sniper fire by ordering air and artillery strikes on villages. As voiced by a US Army officer after the destruction of the Mekong Delta village of Ben Tre, US military philosophy could be: 'We have to destroy the town to save it.'[23]

Notwithstanding this evidence about US war policies, Vietnam Revisionists continue to assign blame for the US failure in Vietnam to misinformed Americans. The culprits are the media and the anti-war movement. Presumably television and influential newspapers turned people against the war. But careful analyses of the national news media demonstrate that, until the Tet Offensive, organisations like the *New York Times* and Columbia Broadcasting System (CBS) accepted the Cold War consensus and the doctrine of containment.[24] Prior to 1968, investigative journalism did not characterise the US media; critical accounts of administration policies conveyed significant congressional attacks or the critiques of distinguished Americans. These early Vietnam critics included Senator J. William Fulbright, the chair of the Senate Foreign Relations Committee, George Kennan, the author of the containment doctrine, and decorated veterans of previous conflicts, such as General David Shoup, General Matthew Ridgeway, and General John Gavin.[25] Only after the communists shattered the confident assessments of President Johnson and General Westmoreland did the national media begin to question sharply the US intervention in Vietnam. As one scholar concluded, 'Walter Cronkite [the chief CBS correspondent] did not drive LBJ from the White House. General Giap did.'[26]

The communists undoubtedly counted on discontent in the United States. Henry Kissinger lamented that North Vietnamese officials, during the Paris peace negotiations, constantly flaunted the strength of the anti-war movement in the United States. But opponents of the war never constituted a majority of American society. And anti-war demonstrators were more unpopular with the American public than was the war itself. Indeed, with his 'silent majority' strategy, Richard Nixon effectively portrayed opponents of the war as unreasonable and even disloyal.[27]

Whatever the merits of the anti-war movement, the question must be asked: did it constrain the military effort in Vietnam? With the War Powers Act of 1973, Congress limited the President's authority to wage war. But that legislation came after the Paris Accords, in reaction to the debacle in Vietnam. During the conflict, anti-war measures, such as the Hatfield-McGovern amendment of 1970 which would have required the withdrawal of US forces from Vietnam by the end of 1971, never secured a congressional majority. In a comprehensive study of the anti-war movement's effect on the executive branch, Melvin Small found that in only one case did the White House discard a military option because it feared a domestic uproar. In part, President Nixon abandoned 'DUCK HOOK' in late 1969, because he was advised by Secretary of State William Rogers and Secretary of Defense Melvin Laird that airstrikes on North Vietnamese cities and a naval blockade would inflame the anti-war movement.[28] But, as heretofore outlined, Henry Kissinger's study group questioned whether an intensification of the war would convince Hanoi to accept US terms.

Vietnam Revisionism falters because, as historian George Herring has pointed out, it ignores the complex realities of the war. For one, it underestimates the dedication of the North Vietnamese and their southern brethren, the *Viet Cong*. At the beginning of the war, US leaders brimmed with optimism. In 1965, Secretary of State Dean Rusk told President Johnson: 'I have a feeling that the other side is not that tough'. In January 1966, the Joint Chiefs of Staff argued that doubters like Secretary of Defense Robert MacNamara exaggerated the 'will of the Hanoi leaders to continue a struggle which they realise they cannot win in the face of progressively greater destruction of their country'. As the war dragged on, a frustrated President Nixon growled that 'a fourth-rate power must have a breaking point'.[29]

In fact, the communists were fanatics, dedicated to their 'sacred mission' of revolutionary nationalism. Moreover, the Vietnamese were patriotic and xenophobic, with a two thousand-year tradition of resisting outsiders. In 1946, Ho Chi Minh had warned the French colonialists that the Vietnamese were prepared to suffer ten casualties for every one they inflicted on the French. During the American phase of their struggle, they bore similar losses, with the North Vietnam losing up to three per cent of its population. With such fortitude, the communists could dismiss as 'irrelevant' Colonel Summers' undeniable observation that US forces defeated the communists in every major encounter. The communists probably were, as one US general confided in correspondent Stanley Karnow, 'the best enemy we have faced in our history'.[30]

Theoretically, the United States could have waged total war against North Vietnam. But, again, how bombing the dykes and flooding the rice fields of the Red River Delta would end insurgency in the South is unclear. In any case, those who advocated annihilating the North Vietnamese ignored a second fundamental reality of the war, its international dimension. The Vietnamese communists had the critical assistance of the rival communist giants, the Soviet Union and the People's Republic of China. Neither nation would have permitted the extinction of North Vietnam. Beyond guaranteeing North Vietnam's survival, both nations frustrated the US bombing campaigns by providing military material to Hanoi. Direct attacks on these modes of supply – Soviet ships, Chinese railways and roads – risked escalating a regional conflict into a global, nuclear confrontation.

In historical enquiry, answers and interpretations are partially shaped by questions asked. The Vietnam Revisionists want scholars to focus on why the United States did not win the war, instead of exploring why the Vietnamese communists withstood the American onslaught. Another question the revisionists might profitably ask is why the South Vietnamese government was unable to defend the country. The *Viet Cong* were winning the war, when, in March of 1965, President Johnson committed US ground troops to Vietnam. In the view of presidential adviser George Ball, the designated critic of US policy, Johnson's decision was 'the quintessence of black humour'. The President and his men had turned 'logic upside down' and 'interpreted the crumbling of the South Vietnamese government, the increasing success of the *Viet Cong* guerrillas, and a series of defeats of South Vietnamese units in the field not as one might expect – persuasive evidence that we should cut our losses and get out – but rather as proving that we must promptly begin bombing to stiffen the resolve of the South Vietnamese government'.[11] Johnson and his advisors accepted the containment doctrine and the domino theory as premisses, not analytic tools. They asked only how South Vietnam could be saved, instead of debating why and whether its independence could be secured.

Through programmes such as 'Vietnamisation', the United States built on paper a powerful South Vietnamese fighting force. By 1973, one million southerners were in uniform, and they had the fourth largest combat airforce in the world. Left alone, however, they proved no match for the communists. In 1974, desertions reached a record high of 240,000. When it launched its final offensive in 1975, Hanoi planned for two years of hard fighting. The South Vietnamese held out for 55 days.[12] The collapse of South Vietnam underscored what Richard Nixon had privately admitted earlier, when he called the Saigon regime 'the weak link in the chain'. Nixon added: 'The real problem is that the enemy is willing to sacrifice in order to win, while the South Vietnamese simply aren't willing to pay that much of a price in order to avoid losing.'[13]

To be sure, as the revisionists have argued, Congress sharply curtailed military assistance to the South after the Paris Accords of 1973. Military aid fell from $2.3 billion in 1973 to $1 billion in 1974. During the last months of the war, the communists probably received more outside assistance than did Saigon. But to charge, in Nixon's words, that the United States 'lost the peace' by abandoning an ally is to forget what happened in the years immediately preceding the Paris Accords.[34] In February 1971, North Vietnamese regulars routed the South Vietnamese when the southerners attacked them in their Laotian sanctuaries. The world saw photographs of South Vietnamese soldiers, desperate to flee the battlefield, clinging to the skids of helicopters. In the spring of 1972, North Vietnam launched its own conventional offensive in the South. Only a massive US bombing operation, code-named 'LINEBACKER', saved the South Vietnamese army from collapse. Former President Nixon apparently forgot that these defeats forced the United States to make critical concessions at the Paris negotiations. The North won a cease-fire-in-place; they were permitted in 1973 to keep approximately 150,000 troops in the South.[35]

Why Saigon could not motivate its population to fight with the same determination as the communists remains a critical issue. Perhaps a majority of southerners were anti-communist. But scholars have observed that the government needed to give the army and the people something to fight for, usually meaning wide-reaching land and tax reform. The revisionists agree and are troubled by the government's short-sighted resistance to change. Guenter Lewy has even suggested that the United States should have played the role of 'good colonialist' and ordered the redistribution of land to peasants.[36] Colonialism and 'nation-building', the avowed aim of the United States in Vietnam, seem, however, mildly contradictory.

To create an independent, non-communist South Vietnam was beyond the power of the United States. As long as the United States deployed air and ground forces to the region, the war would not be lost. But Vietnam made clear the dangers of a globalist foreign policy and the need to distinguish between vital and peripheral interests. However rich and powerful, the United States was incapable of imposing its will upon a determined, albeit much weaker, foe. The American experience in Vietnam demonstrated the limits of national power in an age of international diversity and nuclear weaponry.

NOTES

1   George C. Herring, *America's Longest War: the United States and Vietnam, 1950–1975* (2nd edn, New York 1986); Stanley Karnow, *Vietnam: a History* (New York 1983); Philip Caputo, *A Rumor of War* (New York 1977); Robert Mason, *Chickenhawk* (New York 1983); William Broyles, Jr, *Brothers in Arms; A Journey from War to Peace* (New York 1986).

2   Larry Heinemann, *Paco's Story* (New York 1987); James Webb, *Fields of Fire* (New York 1978); David Rabe, *The Basic Training of Pavlo Hummel* and *Sticks and Bones* (New York 1978); for plays by Cole and others, see *Coming to Terms: American Plays and the Vietnam War*, with an introduction by James Reston, Jr (New York 1985).

3   For an analysis of the ways young men in the United States avoided being sent into combat, see Lawrence A. Baskir and William A. Strauss, *Chance and Circumstance* (New York 1978), pp. 5–10. The authors estimate that, of the 27 million men who came of conscription age during the Vietnam war era, only 6 per cent actually saw combat.

4   Walter Lippmann, *The Cold War: a Study in US Foreign Policy* (New York 1947), pp. 11–28; 52–62.

5   Ford quoted in General Service Administration, *Public Papers of the President: Gerald R. Ford, 1975*, vol. i (Washington 1977), p. 243.

6   Don Oldenburg, 'The Blackboard and the Jungle', *Washington Post*, 30 January 1987, p. C5; William L. Griffen and John Marciano, *Teaching the Vietnam War* (Montclair, NJ 1979), pp. 15–51; Frances Fitzgerald, *America Revised: History Schoolbooks in the Twentieth Century* (New York 1980).

7   US Grant Sharp, *Strategy for Defeat* (San Rafael, Calif. 1978); William Westmoreland, *A Soldier Reports* (Garden City, NY 1976); Peter Braestrup, *Big Story!* (2 vols, Boulder, Colo. 1977); Norman Podhoretz, *Why We Were in Vietnam* (New York 1982); Richard Nixon, *No More Vietnams* (New York 1985).

8   Guenter Lewy, *America in Vietnam* (New York 1978); Harry G. Summers, Jr, *On Strategy: a Critical Analysis of the Vietnam War* (New York 1984); Bruce Palmer, Jr, *The 25-Year War: America's Military Role in Vietnam* (Lexington, Ky. 1984). See also Peter Braestrup (ed.), 'Vietnam as the Past', *The Wilson Quarterly*, vii (Summer 1983), pp. 95–139; Fox Butterfield, 'The New Vietnam Scholarship'. *New York Times Magazine*, 13 February 1983, pp. 26–33, 45–55.

9   Guenter Lewy, 'Vietnam: New Light on the Question of American Guilt', *Commentary*, lxv (February 1978), pp. 49.

10   Walter LaFeber. 'The Last War, the Next War, and the New Revisionists', *Democracy*, i (January 1981), pp. 93–103; see also Thomas G. Paterson, 'Historical Memory and Illusive Victories: Vietnam and Central America', *Diplomatic History*, vii (Winter 1988), pp. 1–18.

11   Herring, *America's Longest War*, p. 256; Karnow, *Vietnam*, pp. 24–6; *Vietnam: a Television History*, produced by Richard Ellison (WGBH Television, Boston 1984), part 13.

12   Thomas W. Netter, 'Vietnam: a New Fear for the Land', *New York Times*, 21 May 1985. p. 17, Broyles, *Brothers in Arms*, pp. 217–19; Herring, *America's Longest War*, p. 151; Fred Wilcox, *Waiting for an Army to Die: the Tragedy of Agent Orange* (New York 1983).

13   David Halberstam, *The Best and the Brightest* (New York 1972), pp. 622–731; Herring, *America's Longest War*, pp. 124–85.

14   Mark Clodfelter, *The Limits of Air Power: the American Bombing of North Vietnam* (New York 1989), pp. 25–38.

15   Johnson quoted in Herring, *America's Longest War*, p. 178.

16   Clark M. Clifford, 'A Vietnam Reappraisal: the Personal History of One Man's View

and How It Evolved', *Foreign Affairs*, xlvii (July 1969), pp. 609–13.

    Loren Baritz, *Backfire: a History of How American Culture Led Us into Vietnam and Made Us Fight the Way We Did* (New York 1985), pp. 259–71; Frances Fitzgerald, 'The Vietnam Numbers Game', *Nation*, 26 June 1982, pp. 776–8.

18 Summers, *On Strategy*, pp. 90–5, 169–73.

19 Seymour M. Hersh, *The Price of Power: Kissinger in the Nixon White House* (New York 1983), pp. 120–34.

20 Arnold Isaacs, *Without Honor: Defeat in Vietnam and Cambodia* (Baltimore 1983), pp. 188– 240; William Shawcross, *Sideshow: Kissinger, Nixon and the Destruction of Cambodia* (New York 1979).

21 Gary R. Hess, 'The Military Perspective on Strategy in Vietnam: Harry Summer's *On Strategy* and Bruce Palmer's *The 25-Year War*', *Diplomatic History*, x (Winter 1986), pp. 104–5; *New York Times*, 13 July 1989, p. A20.

22 Broyles, *Brothers in Arms*, pp. 242–3.

23 Jeffery Race, *War Comes to Long An: Revolutionary Conflict in a Vietnamese Province* (Berkeley 1972), pp. 230–6; Herring, *America's Longest War*, p. 192.

24 For a representative sample of the media's attitude toward the war, see Robert J. MacMahon (ed.), *Major Problems in the History of the Vietnam War* (Lexington, Mass. 1990), pp. 516–27. See also Daniel C. Hallin, *The Uncensored War: the Media and Vietnam* (New York 1986), pp. 105–67; Michael Mandelbaum, 'Vietnam: the Television War', *Daedalus*, lxi (Fall 1982), pp. 157–69.

25 Melvin Small, *Johnson, Nixon, and the Doves* (New Brunswick, NJ 1988), pp. 75–81; Bob Buzzano, 'The American Military's Rationale against the Vietnam War', *Political Science Quarterly*, ci, no. 4 (1986), pp. 559–76.

26 Hallin, *The Uncensored War*, pp. 167–71, 211–15.

27 Braestrup (ed.), 'Vietnam as the Past', pp. 106–7; William L. Lunch and Peter W. Sperlich, 'American Public Opinion and the War in Vietnam', *Western Political Quarterly*, xxxii (March 1979), pp. 21–44; John E. Mueller, *War, Presidents, and Public Opinion* (New York 1973).

28 Small, *Johnson, Nixon, and the Doves*, pp. 180–7.

29 Rusk and Joint Chiefs quoted in Clodfelter, *The Limits of Air Power*, p. 31; Nixon quoted in Herring, *America's Longest War*, p. 224; Herring, 'The War in Vietnam', in Robert A. Divine (ed.), *Exploring the Johnson Years* (Austin 1981).

30 Karnow, *Vietnam*, pp. 15–21; James P. Harrison, *The Endless War: Vietnam's Struggle for Independence* (New York 1988), pp. 133–68; Summers quoted in Richard K. Betts, 'Misadventure Revisited', *The Wilson Quarterly*, vii (Summer 1983), p. 99.

31 Ball quoted in Larry Berman, *Planning a Tragedy: the Americanization of the War in Vietnam* (New York 1982), pp. 45, 130–2.

32 Herring, *America's Longest War*, pp. 262–7; Herring, '"Peoples Quite Apart": Americans, South Vietnamese, and the War in Vietnam', *Diplomatic History*, xiv (Winter 1990), pp. 1–23.

33 Nixon quoted in Karnow, *Vietnam*, p. 642.

34 Nixon, *No More Vietnam*, pp. 165–211.

35 Karnow, *Vietnam*, p. 648; Herring, *America's Longest War*, pp. 240–50.

36 Lewy, *America in Vietnam*, p. 440; Samuel L. Popkin, *The Rational Peasant* (Berkeley 1979).

INDEX

Acheson, Dean, 212, 214–15
Ainger, Arthur Campbell, 137
Archer, Colonel Liam, 169
Attlee, Clement, 211, 215, 224
Auchinleck, General Sir Claude, 2
Austen, Jane, 98

Ball, George, 235
Benedict XI, Pope, 18
Bevin, Ernest, 212, 213, 224
Briggs, E.T., 213–14
Blumentritt, General, 178
Bock, General von, 175
Boland, General Harry, 166
Bonaparte, Napoleon, 3, 4, 7, 68, 80–1, 86–98
Brailsford, H.N., 32, 33, 45
Bradley, General Omar, 220
Braestrup, Peter, 229
Brauchitsch, General von, 175
Brewer, John, 68
Brooke, General Sir Alan, 182
Brown, Terence, 136
Broyles, William, 227, 232
Bruce, Edward, 8, 16, 18, 20
Bruce, Robert, 16
Bryan, Colonel Dan, 160, 163, 167, 169, 171
Burke, Edmund, 7, 71
Bush, George, 230

Cambacérès, J.J., 97
Capp, Bernard, 30–1
Caputo, Philip, 227
Catholic Committee, 66
Catholics: and the British army, 66–81; conflict with Royal troops and Protestants in Languedoc, 58, 65; emancipation of, 81; in Irish Militia, 72–3, 76; legal protection of,

in Canada, 70; and remembrance ceremonies in Northern Ireland, 150–1; Catholic Relief Acts, 71, 73
cavalry, greater tendency to mutiny than infantry, 49
Central Intelligence Agency (CIA), 172, 174 n.45, 232
Charles I, King, 31–3
Churchill, Winston, 2, 7, 219
civil-military relations: in France, 56–65, 121, 124–6; in Ireland, 158–72
Clercq, Staf de, 180–1
Clifford, Clark, 231
Clodfelter, Major Mark, 231
Cole, Tom, 227
Collins, Michael, 162, 163
communists: Chinese, 211–25 passim; in Ireland, 165, 166, 170; Vietnamese, 229, 231–6
Conor, William, 143–4
Cosgrove, William, 145
Craig, Sir James, 150, 151
Cromwell, Oliver, 40, 41, 43
Cronkite, Walter, 233

David, Jacques-Louis, 86
Davies, Godfrey, 44
Dempsey, Lotta, 197
de Valéra, Eamon: agrees to attend opening of Irish National War Memorial, 146, 153; discussions with republicans, 167; reported to have received funds from Mexico, 165; in power, 160, 161, 169
Devlin, Joe, 148
Dickinson, P.G.M., 68
Donegan, Patrick, 158
d'Oppède, Baron, 62–4
dress: wide variety of leg- and footwear worn

Index